Operations Research and Simulation in Healthcare

Malek Masmoudi • Bassem Jarboui • Patrick Siarry
Editors

Operations Research and Simulation in Healthcare

 Springer

Editors
Malek Masmoudi
Faculty of Sciences and Technologies
University of Jean Monnet Saint-Etienne
Roanne, France

Bassem Jarboui
Department of Business
Higher Colleges of Technology
Abu Dhabi, United Arab Emirates

Patrick Siarry
Université Paris-Est Créteil Val-de-Ma
Paris, France

ISBN 978-3-030-45225-4   ISBN 978-3-030-45223-0   (eBook)
https://doi.org/10.1007/978-3-030-45223-0

This Springer imprint is published by the registered company Springer Nature Switzerland AG.
The registered company address is: Gewerbestrasse 11, 6330 Cham, Switzerland

Preface

The healthcare system is facing unprecedented challenges and particularly serious economic pressure. Promising alternatives to the traditional healthcare system are developed to improve the quality of service and reduce cost. Researchers in the operations research and artificial intelligence community are involved to provide efficient healthcare decisional support systems to help healthcare engineers and managers to make optimal/efficient/robust decisions in dynamic and uncertain environments. How to make changes in order to improve the whole healthcare system performance becomes the main issue of researchers from the technical, organizational, and decisional points of view. The need for intelligent decision-making techniques in healthcare domain is growing. The domain is complex and very rich in terms of scientific niches which attract researchers in both the operations research (OR) and artificial intelligence (AI) disciplines.

Optimization and simulation techniques have been widely considered in the literature to deal with various decision-making problems in healthcare sector. This book "Operations Research and Simulation in Healthcare" provides recent studies and works in healthcare management and engineering using optimization and simulation methods and techniques. It will focus on mainly exposing the readers to the cutting-edge research and applications that are going on across the domain of healthcare management and engineering where optimization and simulation can be and have been successfully employed.

Need for a Book on the Proposed Topics

To the best of our knowledge, there is no book aiming precisely at regrouping optimization and simulation techniques for healthcare decision-making problems, and the number of book chapters dedicated to this subject is tiny. However, this topic is highly relevant today and interests many researchers, which explains the high number of journal papers and international conference papers dedicated to this subject.

This book "Operations Research and Simulation in Healthcare" provides readers with optimization and simulation tools for solving healthcare decision-making problems. It explains a wealth of both basic and advanced concepts of optimization and simulation for the management of emergency service, patient flows, logistics, home care service, operating theater scheduling, wards, human resources, etc. The chapters include several points of view and relevant case studies.

Organization of This Book

This book is organized into 10 chapters. A brief description of each chapter is given below.

Chapter 1 entitled "A Two-Dimensional Categorization Scheme for Simulation-/Optimization-Based Decision Support in Hospitals Applied to Overall Bed Management in Interdependent Wards Under Flexibility" by T. Stoeck and T. Mellouli provides a categorization scheme of resource-related decision problems based on the scope of planning in terms of levels of organizational and functional unit. A simulation study is provided based on the yearly data of a hospital in Germany. Flexibility in resource management is evaluated in combination with the clustering of wards based on their similarities. A case study of hierarchical optimization efficiently integrates daily planning of patient pathways as an input/instruction for a detailed hourly optimization of treatments. As a general result, it is shown that thinking about flexibility can turn the disadvantages from interdependencies in a capsuled way to advantages in a broader context.

Chapter 2 entitled "A Heuristic Based on the Hungarian Method for the Patient Admission Scheduling Problem" by R. Borchani, M. Masmoudi, B. Jarboui, and P. Siarry tackles the operational level of hospital admission scheduling problem. This combinatorial optimization problem consists in assigning patients to specific beds in specialized departments during a predefined length of stays, subject to satisfying their medical requirements as well as their desires. To address this problem, the authors present a heuristic based on the Hungarian method. The experimentation on benchmark instances shows that the proposed method produces competitive results in comparison with existing works in the literature.

Chapter 3 entitled "A Bi-objective Algorithm for Robust Operating Theatre Scheduling" by T. Chaari and I. Omezine provides a scenario-based optimization approach for the operating theater scheduling problem. A robust bi-objective evaluation function is defined in order to obtain a robust and effective solution that is slightly sensitive to surgical and recovery durations uncertainty. A simulation model is provided to validate the solving approach and the quality of the robustness. Experimental results show a tradeoff between effectiveness and robustness for various degrees of uncertainties.

Chapter 4 entitled "Cost Function Networks to Solve Large Computational Protein Design Problems" by D. Allouche, S. Barbe, S. de Givry, G. Katsirelos, Y. Lebbah, S. Loudni, A. Ouali, T. Schiex, D. Simoncini, and M. Zytnicki deals with the computational protein design (CPD). The challenge of identifying a protein that performs a given task is defined as the combinatorial optimization of a complex energy function over amino acid sequences. The CPD problem can be formulated as a binary cost function network (CFN). The efficiency of the CFN approach is shown on a set of real CPD instances built in collaboration with structural biologists. For very large CPD instances, the CFN approach has been incorporated inside a (parallel) variable neighborhood search method which improves the results and provides optimal solutions for several instances.

Chapter 5 entitled "Modeling and Simulation in the Dialysis Center of the Hedi Chaker Hospital" by I. Jridi, B. Jerbi, and H. Kammoun provides a comprehensive discrete event simulation (DES) model that includes the different stages which a patient may go through in the renal unit within the Hedi Chaker University Hospital of Sfax, Tunisia. The stages start with the patient arrival and finish either with a successful kidney transplant or with his/her death. Switching a number of times between modalities of treatment is really possible and presented in the proposed model. Results show that, in Tunisia, the number of patients undergoing hemodialysis, peritoneal dialysis, or kidney transplant is steadily growing, ensuring in turn the increasing demand for treatment, experimental staff, and materials.

Chapter 6 entitled "Toward a Proactive and Reactive Simulation-Based Emergency Department Control System to Cope with Strain Situations" by S. Chaabane and F. Kadri deals with emergency departments (EDs) organization and improving the level of resilience (proactive and reactive management). Based on the resilience framework, the objective of this chapter is to propose an approach that allows us to improve the EDs' proactive capacities regarding inpatient flow management by combining multivariate forecasting methods and the discrete event simulation (DES) approach. A case study is considered to analyze and evaluate the proactive management of patient flows in the Pediatric Emergency Department (PED) at Lille Regional Hospital Center, France.

Chapter 7 entitled "A Decentralized Approach to the Home Healthcare Problem" by B. Issaoui, A. Mjirda, I. Zidi, and K. Ghédira deals with the daily caregivers' tours while minimizing the travel costs and maximizing the planned services which addresses potential conflicts. A multi-agent approach using three heuristics is provided and tested on a set of benchmarks from the literature. Numerical results show the efficiency of our approach and prove that the policy based on remaining visits gives better results in terms of resolving detected conflicts.

Chapter 8 entitled "Transporting Wounded People and Assignment to Hospitals During Crises" by M. Khorbatly, H. Dkhil, H. Alabboud, and A. Yassine deals with the crisis management supply chain, more precisely minimizing the total evacuation transportation time for wounded people in the case of disasters in order to minimize the human loss. Emergency evacuation planning involves the planning of the required resources assignment and timely vehicle scheduling. To achieve this goal, the authors treat the Integrated Problem of Ambulance Scheduling and Resources Assignment (IPASRA) and formulate it as a linear model, furthermore they provide a novel hybrid algorithm based on Tabu search (TS) and the greedy randomized adaptive search procedure (GRASP).

Chapter 9 entitled "Carbon Footprints in Emergency Departments: A Simulation-Optimization Analysis" by M. Vali, K. Salimifard, and T. Chaussalet proposes the application of simulation techniques to model, analyze, and reduce carbon footprint in hospitals. The stages for doing this are as follows: (1) the treatment processes in the emergency department, (2) sources of carbon production in the emergency department, focusing on equipment, (3) developing a simulation model to capture the treatment processes and the production of carbon, (4) scenarios and outputs analysis, and (5) conclusion and directions for future research.

Chapter 10 entitled "The Effect of Risks on Discrete Event Simulation in Healthcare Systems" by M. Poursoltan, M. Masmoudi, and M. Kaba Traore deals with discrete event simulation (DES) in healthcare for decision-making at the managerial level, i.e. allocation of material, staff, and financial resources, taking into account the inevitable risks related to process variability, uncertainty, scale, and interconnections that are critical in real healthcare systems. Simulation is able to capture all these factors. This chapter draws on the DES project in healthcare and discusses the agenda focusing on (1) a characterization of the DES project and the simulation process, (2) risks applying the DES project in healthcare, (3) applicable risk analysis methods, and (4) recommendations for successful implementation. The chapter contains a case study of an emergency department in Iran.

Audience

This book is valuable for researchers, master's, and Ph.D. students in departments of computer science, information technology, industrial engineering, and applied mathematics, and in particular for those engaged with OR topics in the healthcare domain.

Roanne, France   Malek Masmoudi
Abu Dhabi, United Arab Emirates   Bassem Jarboui
Paris, France   Patrick Siarry

Contents

Contributors

Hassan Al Abboud Laboratory of Informatics and Applications LIA, Doctoral School in Science and Technology DSST, Lebanese University, Tripoli, Lebanon

David Allouche INRA MIAT, UR 875, University of Toulouse, Toulouse, France

Sophie Barbe LISBP, INSA, UMR INRA 792/CNRS 5504, Toulouse, France

Rahma Borchani Faculty of Economics and Management of Sfax, Laboratory of Modeling and Optimization for Decisional, Industrial and Logistic Systems (MODILS), University of Sfax, Sfax, Tunisia

Sondes Chaabane CNRS, UMR 8201 - LAMIH-Laboratoire d'Automatique de Mécanique et d'Informatique Industrielles et Humaines, Université Polytechnique Hauts-de-France, Valenciennes, France

Tarek Chaari Management Information System Department, College of Business Administration, Taibah University, Madina, Kingdom of Saudi Arabia

Thierry Chaussalet Computer Science and Engineering Department, Westminster University, London, UK

Simon De Givry INRA MIAT, University of Toulouse, Toulouse, France

Hamdi Dkhil LMAH, Normandie University, Unihavre, Le Havre, France

Khaled Ghédira Université Centrale de Tunis, Honoris United Universities Tunis, Tunisia

Brahim Issaoui Complex Outstanding Systems Modelling Optimization and Supervision (COSMOS), Strategie d'Optimisation et Informatique intelligentE (SOIE), National School of Computer Science (ENSI), Manouba, Tunisia

Bassem Jarboui Higher Colleges of Technology, Abu Dhabi, United Arab Emirates

Badreddine Jerbi College of Business and Economics, Qassim University, Al Qassim, Kingdom of Saudi Arabia

Ichraf Jridi College of Business Administration, Taibah University, Medina, Kingdom of Saudi Arabia

Farid Kadri Analytics and Big Data department, 1025-Aeroline PCS Agence, Sopra Steria Group, Colomiers, France

Hichem Kamoun Faculty of Economics and Management, University of Sfax, Sfax, Tunisia

George Katsirelos INRA, AgroParisTech, UMR MIA-Paris, University of Paris-Saclay, Paris, France

Mohamad Khorbatly Institut Supérieur des Sciences Appliquées et Economiques, ISSAE-CNAM University, Tripoli, Lebanon

Yahia Lebbah LITIO, University of Oran 1, Oran, Algeria

Samir Loudni GREYC, UMR 6072-CNRS, University of Caen Normandy, Caen, France

Malek Masmoudi University of Jean Monnet Saint-Etienne, Sciences and Technologies Faculty, University of Lyon, Saint-Etienne, France

Taïeb Mellouli Department of Business Information Systems and Operation Research, Martin-Luther-University Halle-Wittenberg, Halle, Germany

Anis Mjirda Faculty of Economic Science and Management, University of Sfax, Sfax, Tunisia

Imen Omezine College of Business Administration, Management Information System Department, Taibah University, Madina, Saudi Arabia

Abdelkader Ouali ORPAILLEUR Team, LORIA, INRIA - Nancy, Vandœuvre-lès-Nancy, France

Milad Poursoltan IMS UMR 5218 laboratory, University of Bordeaux, Talence, France

Khodakaram Salimifard Department of Industrial Management, Persian Gulf University, Bushehr, Iran

Thomas Schiex INRA MIAT, University of Toulouse, Toulouse, France

Patrick Siarry University of Paris-Est Creteil, Paris, France

David Simoncini IRIT, UMR 5505-CNRS, University of Toulouse, Toulouse, France

Thomas Stoeck Department of Business Information Systems and Operation Research, Martin Luther University Halle-Wittenberg, Halle, Germany

Mamadou Kaba Traore IMS UMR 5218 laboratory, University of Bordeaux, Talence, France

Masoumeh Vali Department of Industrial Management, Persian Gulf University, Bushehr, Iran

Adnan Yassine Institut Supérieur d'Etudes Logistiques, Normandie University, Unihavre, Le Havre, France

Issam Zidi Complex Outstanding Systems Modelling Optimization and Supervision (COSMOS), Strategie d'Optimisation et Informatique intelligentE (SOIE), National School of Computer Science (ENSI), Manouba, Tunisia

Matthias Zytnicki INRA MIAT, University of Toulouse, Toulouse, France

Acronyms

AD	Admission date
APD	Automated peritoneal dialysis
ACPR	Average cost of patient room assignment
ALOS	Average length of stay
BTD	Backtracking with tree decomposition
CF	Carbon footprint
CPR	Cardiopulmonary resuscitation
CSP	Constraint satisfaction problem
CAPD	Continuous ambulatory peritoneal dialysis
CPs	Controllable parameters
CPD	Computational protein design
CFN	Cost function network
DEE	Dead-end elimination
DSS	Decision support systems
DFBB	Depth-first branch-and-bound
DS	Department specialism
DD	Discharge date
DES	Discrete event simulation
DPM	Disease progression modeling
ED	Emergency department
ED-CS	Emergency departments control system
ESRF	End-stage renal failure
EPT	Equivalence preserving transformation
ETA	Event tree analysis
EDAC	Existential directional arc consistency
FMEA	Failure mode and effect analysis
F&O	Fix-and-optimize
F&R	Fix-and-relax
FJSP	Flexible job-shop scheduling
G-DRG	German diagnosis-related groups
GMEC	Global minimum energy conformation

GLA	Greedy look-ahead heuristic
GRASP	Greedy randomized adaptive search procedure
GHGs	Greenhouse gases
HS	Harmony search
HBM	Health behavior modeling
HCSO	Health and care systems operation
HD	Hemodialysis
HHC	Home health care
HHCP	Home health care problem
HHCS	Home health care service
HA	Hybrid algorithm
HBFS	Hybrid best-first search
IPASRA	Integrated problem of ambulance scheduling and resource assignment
ICU	Intensive care unit
KPMs	Key performance measures
K–M	Kuhn–Munkres algorithm
LAHC	Late acceptance hill climbing
LOS	Length of stay
LDS	Limited discrepancy search
01LP	0/1 Linear programming
LB	Lower bound
MRI	Magnetic resonance imaging
MIP	Mixed integer programming
MILP	Mixed integer linear programming
MAS	Multi-agent system
MOOP	Multi-objective optimization problem
MOGA	Multi-objective genetic algorithm
MCDM	Multi-criteria decision-making
m-TSPTW	Multiple traveling salesman problem with time windows
RFN	Necessary room features
N.S	No solution
NRP	Number of remaining patients
OR	Operating room
OT	Operating theater
OEE	Overall equipment effectiveness
O.M	Out of memory
OR	Operations research
PA	Patient age
PAS	Patient admission scheduling
PASU	Patient admission scheduling under uncertainty
PBA	Patient bed allocation
PG	Patient gender
PRC	Patient-room cost
PS	Patients specialism

PED	Pediatric emergency department
PD	Peritoneal dialysis
PH	Planning horizon
RFD	Preferred room features
PW	Primary waiting time
PCA	Principal component analysis
PMI	Project Management Institute
PDB	Protein Data Bank
RRT	Renal replacement therapy
RT	Renal transplant
PR	Room properties
RC	Room capacity
RP	Room capacity preference
RS	Room specialism
SM	Screening modeling
SHU	Short-term hospitalization unit
SA	Simulated annealing
SOOP	Single-objective optimization problem
SPC	Statistical process control
TS	Tabu search
TWL	Transplant waiting list
TR	Transfer
UPs	Uncontrollable parameters
UDGVNS	Unified decomposition guided variable neighborhood search
VNS	Variable neighborhood search
V.H.A	Value given by hybrid algorithm
V.C	Value given by CPLEX
VRPSDTW	Vehicle routing problem with discrete split delivery and time windows
VRP	Vehicle routing problem
VRPTW	Vehicle routing problem with time windows
V&V	Verification and validation
VIKOR	Visekriterijumska optimizacija i kompromisno resenje
VNS	Variable neighborhood search
WCSP	Weighted constraint satisfaction problem
WHO	World Health Organization
WIP	Work in process

Chapter 1
A Two-Dimensional Categorization Scheme for Simulation/Optimization-Based Decision Support in Hospitals Applied to Overall Bed Management in Interdependent Wards Under Flexibility

Thomas Stoeck and Taïeb Mellouli

Abstract Hospitals are facing unprecedented challenges and economic pressure and therefore forced to find new ways to make their processes more effective and use their resources more efficiently. Our research is based on evaluating flexible usage of narrow resources across departments and wards without resort to high investment. Overall bed management in hospitals is considered for this evaluation.

Focusing on decision support in hospitals, six types of interdependencies concerning, e.g., resource capacities and processes, are discussed. To cope with such complexities, we introduce abstraction levels of demand side around patient pathways as well as supply side concerning resources. Based on this differentiation, we propose a new two-dimensional categorization scheme of resource-related decision problems based on the scope of planning in terms of levels of organizational and functional unit, namely single/multiple pathways, single/multiple wards, and hospital wide, combined with the detailed level of resource planning in terms of granularity for (daily) bed capacities and treatments. An extensive literature review is presented within this scheme in order to link the state-of-the-art to our proposed research.

In simulation studies based on yearly data of a university hospital, we show that some types of interdependencies should be considered in order to get valid results of patient flows. Toward flexibility in resource management, a scenario of temporarily sharing beds of nearby wards showed considerable savings in patient rejection and lengths of stays. This flexibility is evaluated in combination with clustering of wards based on their similarity w.r.t. offered kinds of patient treatments. A case study of hierarchical optimization efficiently integrates daily planning of patient pathways

T. Stoeck (✉) · T. Mellouli
Department of Business Information Systems and Operations Research, Martin-Luther-University Halle-Wittenberg, Halle, Germany
e-mail: thomas.stoeck@wiwi.uni-halle.de; mellouli@wiwi.uni-halle.de

© Springer Nature Switzerland AG 2021
M. Masmoudi et al. (eds.), *Operations Research and Simulation in Healthcare*, https://doi.org/10.1007/978-3-030-45223-0_1

as input/instruction for a detailed hourly optimization of treatments. As a general result, it is shown that thinking about flexibility can turn the disadvantages from interdependencies in a capsuled way to advantages in a broader context.

1.1 Introduction

The healthcare system and hospitals are facing unprecedented challenges and particularly a serious economic pressure. In order to survive in this difficult market, hospitals are forced to find new ways to make their processes more effective and use their existing resources more efficiently. Especially, the hospital-wide bed management has a huge impact on the performance of a hospital. A bed is seen as a central resource of a hospital because it is subsuming all following resources like physicians, nurses, operating rooms, and equipment. Without an appropriate bed for patient admission, none of the other resources can be planned or used. Patient numbers are rising because of the demographic change of the society and higher rates of elderly people, while bed capacity is constant or decreasing.

To face these new challenges, a hospital has at least two opportunities. The first opportunity consists of an efficient use of existing bed capacities, which is sometimes crucial to the economic survival of hospitals. Here, computer-assisted decision support systems may help by simulating scenarios of patient flows under high bed utilization (scenarios with higher elective patients'·admission and endogenous, stochastically determined emergency patients' rates) and by optimization of operations, patient admission, and pathways under highest rate of resource utilization. However, this may be not enough in case of tight/narrow resources. The second but costly alternative action for a hospital is to invest in new beds and infrastructure in order to increase its capacity, which results in very high costs for acquisition and maintenance. Furthermore, staff costs would rise because the number of beds and the amount of needed staff are linked. Due to enormous economic pressure, this is mostly not possible for hospitals.

Our research is motivated by finding new ways to resolve these challenges and difficulties in case of very narrow hospital resources without resort to high hospital investment. We hereby clearly reject the second opportunity as much as possible and enlarge the applicability of the first opportunity by the idea of flexible use of tight resources in a hospital-wide view. We augment the first opportunity (decision support systems by simulation and optimization) with operational flexibility strategies that especially implement flexibility of resource usage respecting all expert rules and constraints governed by hospital organizational and medical needs. As good overall performance of hospitals directly depends on overall good resource planning, we study interdependencies influencing decision support problems and recognize and discuss different kinds of them.

Optimization and simulation constitute a broad and rich scientific area in healthcare research. With a huge body of literature, an effective categorization scheme is important for organizing current and future research, and additionally, for recognizing possible research gaps and interesting research questions. Based on our work on synergies between AI methods and optimization [1], we develop a two-dimensional categorization scheme and classify different decision support areas in the recent scientific literature. As a preparation, we present and discuss six different kinds of interdependencies and abstraction levels of resources and patient pathways.

In the practical part of the chapter, we present two different types of case studies using simulation and optimization and then classify the considered decision support problems within our two-dimensional categorization scheme. The considered case studies propose practically feasible ways to implement flexibility strategies aiming at an effective usage of shared or compatible resources. It is shown that disadvantages of interdependencies can be overcome by such flexibility strategies that can be effectively implemented across different organizational units and problem areas. In a retrospective view considering both theory and practice, flexibility strategies in a hospital-wide bed management are evaluated with respect to findings of the current state of implementation and extended for future research steps.

Outline of Chapter The remainder of the book chapter is organized as follows. In the following second section, decision support in hospitals is considered in an overall planning context considering the complexities of the demand side through patient flows and supply side of bed and resource management. We recognize and discuss six different kinds of interdependencies explaining kinds of these complexities surrounding decision support and planning problems in hospitals. In order to manage these interdependencies, we introduce in the third section abstraction levels both of patient flows together with the scope planning and inner hospital resources together with their detailed level of consideration. Based on this differentiation, we propose a new two-dimensional categorization scheme for resource-related decision problems in hospitals. In this scheme presented in the fourth section, a first dimension considers the scope of planning in terms of levels of organizational and functional unit, namely single/multiple pathways, single/multiple wards, and hospital wide, while the second dimension encompasses the detailed level of resource planning in terms of granularity for daily bed capacities and treatments and their aggregation at a weekly, monthly, and yearly basis. Based on this scheme and an extensive literature search, different problem areas of decision support in hospitals are discussed and classified.

In the practical part of this chapter in the fifth section, we present real-life case studies of overall bed management and treatment planning based on own experiments and evaluations using both simulation techniques [2, 3] and mathematical optimization [4]. Our simulation and optimization components operate on concrete types of patient pathways that can be gained by data and process mining components, including special clinical pathway mining algorithms [5]. These case studies promote a possible effective implementation of flexibility in resource usage in overall hospital decision support. Considered flexibility strategies range from sharing beds of nearby wards to clustering strategies of wards based on similarity of treatments offered by clustered wards. Scenario experimentations and evaluations based on real data of a university hospital in our works [2–4] validate our suggestion that an overall bed management in interdependent wards is crucial for decision support in hospitals.

Furthermore, experimental results show that implementing flexibility in resource usage contributes a winning strategy for hospitals. The crucial importance of flexibility in hospital decision support together with ideas about clustering and quasi-cluster is presented in the first African conference on operations research [6]. In the sixth section, flexibility strategies are considered in their own, and their impact is discussed based on the state of experimentation and evaluation. After showing some research issues in this respect, the chapter is closed by concluding remarks in the seventh section concerning both theoretical and practical findings of our research and development work discussed in this chapter.

1.2 Decision Support in Hospitals with Respect to Overall Planning

Decision support is a broad field, which receives a lot of attention in the scientific literature. In order to discuss hospital decision support and its implication, we describe the basic processes of a hospital with respect to decision support and the possible interdependencies influencing decision support in hospitals will be discussed.

1.2.1 Processes in Hospital Decision Support

In general, decision support through specialized application systems supports the decision maker with models, methods, and problem-related data. Characteristic of decision support systems is the distinctive model and method orientation [7]. For the use in healthcare or in hospitals, an extension of the well-known definition is necessary. Although hospitals are forced to work in a more efficient way due to changing economic conditions, they differ in essential points from classical enterprises. Especially in a hospital, a distinction between core and support processes is made [9]. Unfortunately these processes, as classically presented and classified, are only considered at an organizational point of view that does not reflect the complexity of planning tasks in a hospital. The distinction between core and support processes in [9] neglects the connection between core medical processes and supporting tasks like resource or staff processes from a planning point of view. Because of that we will focus on a missing process consideration in classical approaches by delivering a point of view where planning is a central activity or process in a hospital.

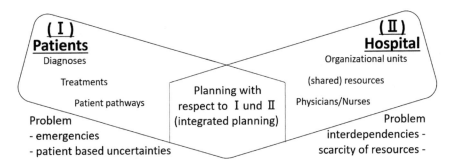

Fig. 1.1 Central planning task in hospitals—I: Demand and II: Supply

In Fig. 1.1 the two main wings of the overall planning perspective of a hospital are shown. In Fig. 1.1 (I) the patient point of view is considered. In general patients are classified via their diagnoses. Based on that, possible treatments are set. Basically a combination of possible treatments is called a patient pathway. The main difficulties arising here are the occurrence of emergencies and patient-based uncertainties like arrival times, different treatment paths and different treatment times. In general wing (I) describes the problem of composing a service, generally represented and identified as a patient pathway, for different patients in order to recreate or to maintain a positive health status. Hereby a first complexity of planning arises by ensuring the combinability of diagnoses, treatments, and other services and the simultaneous consideration of elective and emergency patients.

In Fig. 1.1 (II) the hospital point of view is described. Basically a hospital consists of different organizational units like departments, diagnostic centers, and wards. Furthermore, hospitals have to deal with generally scarce resources, for example, beds and equipment. Additionally, based on the context, shared resources exist, which are used by multiple units at the same time. Moreover, physicians and nurses have to be planned inside wards of a hospital. Besides the scarcity of resources, a main problem within this wing is the existence of interdependencies of different kinds. In this wing a second type of complexity arises by considering the feasibility of a certain patient mix with respect to available resources. In general the main planning task from an operations research perspective for a hospital is located in the intersection between I and II.

Inside this intersection the bed management, as a central connecting decision support task between patient pathways and resources, is located. As stated before, beds are considered as one of the most important resources inside a hospital. Without a proper bed for a patient, no other planning task is possible because every treatment or intervention is linked to a bed for a patient. Basically this means that every resource inside a hospital is subsumed by beds from a planning point of view. Bed and treatment management are strongly interrelated and can be tackled by means of hierarchical optimization [4].

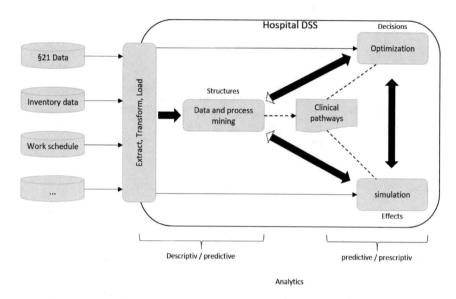

Fig. 1.2 An overall decision support concept for hospitals [8]

Furthermore, all problem areas have a critical demarcation from classical economic decision support. Nearly every decision of a hospital depends on the patient. A patient-centered approach naturally introduces a series of uncertainties into the decision-making process. These uncertainties can be divided into different areas. First, it is difficult to forecast the arrival of patients, especially for emergency patients. In addition, it is difficult to predict the course of treatment of patients [10]. Different diagnoses often result in different treatment paths and, depending on the condition of the patient, different treatment times [11]. These patient-related uncertainties greatly complicate decision-making in a hospital and in healthcare in general. Because of this patient-centered approach, our research team defines a hospital-wide decision support concept based on patient pathways (Fig. 1.2) [8]. Within this concept, optimization is used to perform decisions and simulation studies on the effects of these decisions. Furthermore this concept combines techniques of data/process mining, optimization, and simulation to enable new ways of decision-making.

In addition to the uncertainties, interdependencies between individual organizational units of a hospital have a great influence on the complexity of the decision-making process and provide interesting questions and use cases for researchers and practitioners. Furthermore, we integrate flexibility strategies of resource usage in our experiments. These are evaluated in our case studies presented in Sect. 1.5 and further discussed in Sect. 1.6.

1.2.2 Types of Interdependencies in Hospital Processes and Decisions

Interdependencies generally refer to interactions of at least two actors, processes, or issues. In the current research literature on decision support in hospitals, these interdependencies are not examined and classified in detail. There are a few articles that discuss interdependencies in a hospital. Unfortunately they mostly focus on one kind of interdependencies. So a broad overview about interdependencies is lacking. Furthermore, in most cases researchers state that there are interdependencies and they make assumptions in order to cope with them. Because of that, this section provides a more detailed classification and analysis of possible interdependencies in a hospital. In general, it can be assumed that, depending on the level of abstraction of the decision problem, the nature and influence of interdependencies changes. In many cases, this has a major impact on the modeling of the problem. The more interrelationships are considered, the more complex the problem becomes [12]. In order to discuss interdependencies and their implication for decision problems, we present and propose six main types of interdependencies to their classification inside a hospital (Table 1.1).

Capacity-based interdependencies are characterized by the competition for shared and narrow resources. Basically different patients have a certain demand for the same shared resource like beds or medical equipment [13, 14]. By validating results in our case studies [2, 3] (Sect. 1.5.1), we observed two additional aspects of capacity-based interdependencies respecting room capacity. First, bed capacity in a ward may be shortened by diagnosis and gender mix of patients. Some diagnosis of single patients with communicable diseases may necessitate that a multibed room must be assigned to a single patient. Second, because mixed men/women occupancy is prohibited, a certain gender mix of patients may necessitate that

Table 1.1 Kind of interdependencies in a hospital

No.	Interdependency	Properties
1	Capacity-based	Competition for shared resources
		A decision has a possible impact on further decisions
2	Process-based	Effects of process improvements are investigated
		Effects of local optimization depend on the position in the system
3	Functional-based	Arising from treatment of various diagnosis
		With rising numbers of diagnosis uncertainties increase
4	Staff-based	Employee specialization and hierarchy levels are considered
		Effects on staff planning and pooling
5	Problem-based	Relation between separate decision problems
		On the same or on different hierarchical levels
6	Patient-based	Relationship between elective and emergency patients
		Forecasting of emergencies affects the available capacity for elective patients

not all beds of a ward can be occupied. Additionally this decision has a huge impact on potential treatment capacities. Depending on the diagnosis, different treatments may be necessary. Therefore, a second competition arises by competing for timeslots in specialized units. Therefore, decisions that are good at the bedside level, for example, optimizing the occupancy rate of beds, can have a negative impact on further capacity decision problems on treatment level. Therefore, the capacity bottleneck is pushed just one step further [14]. So optimizing resources for all patient groups at the same time is a complex task [13].

The second type of interdependencies involves the *process-based interdependencies*. These interrelations mainly relate to the effects of process improvements. A hospital can be seen as a so-called downstream system [15]. This means that the flow of patients is primarily directed in one direction. This means that patients enter the hospital via a few facilities, such as a central emergency room, and will be distributed to other facilities. A characteristic of downstream systems is the fact that an isolated view of a domain can have a negative impact on the overall performance of a system. The effects of local optimizations on the overall performance depend on the position of the considered area in the overall system. In particular, process improvements in the main admission facilities lead to problems in the following areas[15].

Functional interdependencies denote the interdependencies that arise from the treatment of various diagnoses or diseases. With a rising number of diseases treated, the uncertainty for the planning process and for capacity supply also increases. Functional units that treat only one disease have little functional interdependence. For such units, standard processes can be developed in order to simplify planning, while units that treat a high number of illnesses need integrated planning solutions to incorporate the interrelationships[16].

The fourth kind of interdependencies is composed of *staff interdependencies* that are dealing with employee specialization and hierarchy levels of workers. Also the effects from staff planning and pooling approaches on other problem areas are covered within this kind of interdependencies [14, 17–19].

The next and mostly neglected but crucial kind of interdependencies covers *problem-based interdependencies*. This kind describes the relation between separate decision tasks. In general combinations of previous interdependencies are possible. For example, when a capacity planning of a ward is performed, the possible consequences for the planning of treatment capacities should be respected. This means that for a particular decision problem the related upstream or downstream problems should be respected in order to achieve results usable in practice. But in a hospital environment it is difficult to characterize upstream and downstream problems. In most cases a decision problem has multiple interdependencies with other decision areas, which have multiple relations on their own.

In connection with the last paragraph a sixth kind of interdependency can be discussed. These kind covers *patient-based interdependencies* between elective patients and emergency patients. From a planning point of view, this is a crucial kind of interdependency on the highest level of planning. The available capacity for elective patients depends on the quality of the forecast of emergency patients. As every planning task in a hospital is correlated to the amounts of patients, therefore it is crucial for hospital decision support to cope with this kind of interdependency.

1.3   Abstraction Levels of Decision Support in Hospitals

This section examines possible abstraction levels of decision support in a hospital, which are necessary to further classify arising problem areas and decision support applications. First of all, the possibilities of abstraction of patient pathways are discussed. Subsequently, resources are presented as a second type of abstraction.

1.3.1   First Abstraction: Patient Pathway and Scope of Planning

According to Bartz (2006), there is no standard definition of patient pathways [20]. From different definitions, the nature and characteristics of a patient's pathway can be discussed. In general, a patient pathway can be understood as an instrument that represents a comprehensive diagnosis-oriented interdisciplinary treatment plan for defined patient groups or a virtual average patient [21], which corresponds to the first column of our two-dimensional scheme. A patient path has two main objectives [22, 23]. On the one hand, qualitative medical goals and, on the other hand, economic efficiency-oriented goals are pursued. These two main goals are usually in conflict with each other, so an efficient solution within the resulting trade-off has to be found as foundation for a good cost management [24]. Pathways can still be understood as an operational version of a guideline [22]. In general a reasonable planning of patient pathways enables the hospital management and medical staff to avoid unnecessary waiting times for patients [25].

The presented definition, however, is aimed more at operational considerations. For this chapter, the definition has to be extended by introducing abstraction levels in the context of patient paths. Regardless of the level of detail, we use the previous definition as an abstract patient pathway (Fig. 1.3a). Further abstraction levels are based on considered organizational units and the number of concurrent or parallel patient pathways. On the second level of abstraction (Fig. 1.3b), all or many abstract patient paths within a ward or functional unit are considered. The special feature is that the different paths compete for shared resources. The third level of abstraction (Fig. 1.4a) considers path groups or multiple paths at an aggregated organizational level, for example, on department or cluster level. The fourth and last level of abstraction (Fig. 1.4b) examines hospital-wide (individual) paths. There are two variants, which are summarized at this level. Path groups or patient-specific paths can be considered.

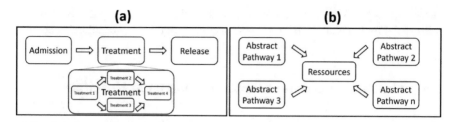

Fig. 1.3 Different abstractions of patient paths at one organizational unit. (**a**) Abstract path. (**b**) Many paths

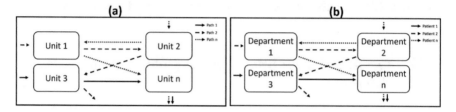

Fig. 1.4 Different abstractions of patient paths at multiple organization units. (**a**) Department-wide paths. (**b**) Hospital-wide paths

1.3.2 Second Abstraction: Resources and Detail Level of Planning

Another level of abstraction to classify planning problems is the subdivision at resource level. Here, a partition based on the degree of detail of the resource is presented. The most detailed planning level is at treatment level. A treatment is a completed activity that is performed to improve or maintain a patient's state of health. The planning deals with the sequence planning and the chronological planning of individual treatments on a granular level.

The next level of planning is the daily planning of beds or treatment capacities. This is a key requirement for the treatment of patients. Occupancy management is involved in nearly every hospital process [26]. In the literature [27] a bed is considered as one of the most important and expensive resources inside a hospital. Good planning and utilization of existing bed capacity is a prerequisite for fast and smooth treatment of patients. Above all, the daily planning of the beds is considered within this planning level.

Schwarz et al. [4] present bed management and treatment scheduling as a hierarchical interdependent problem. In Fig. 1.5 the integrated planning of the first and the second abstraction levels is shown. So the daily planning of beds is done with respect to a possible planning on treatment level. A more detailed discussion of this interrelated decision task is given in Sect. 1.5.2. For a more strategic planning of bed capacities, the next level of abstraction is used. Here, bed capacities or resource

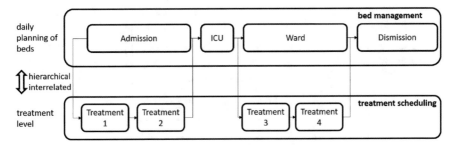

Fig. 1.5 Hierarchical planning of beds and treatments [4]

capacities are planned in an aggregated way over a long period of time. Thus, a strategic planning of capacities can take place here [14].

1.4 Two-Dimensional Categorization Scheme

In order to use the presented abstraction levels to classify problem areas and current research direction, we designed a two-dimensional categorization scheme for simulation- and optimization-based decision support tasks in hospital, which is presented and discussed in Sect. 1.4.1. Within this scheme, a classification of problem areas will be presented in Sect. 1.4.3 and a literature review for certain problem areas shows insights about the current research direction and methods used.

1.4.1 Two-Dimensional Categorization Scheme for Decision Support in Hospitals

As stated in previous sections, decision support via simulation and optimization is a broad research field in various problem areas inside a hospital. A wide range of literature reviews shows that a lot of problem areas inside a hospital exist. In [12] a general overview about using system sciences in health care is given. Based on that, some general problem areas are detected. In [28] different problem definitions regardless of their organizational unit are shown. Additionally two other surveys [29, 30] try to classify problem areas regarding their organizational scope of application. In addition to these general reviews there are several specialized literature reviews that cover different scopes. This scope ranges from operating room scheduling [31] over capacity management [32] to capacity planning [33]. Furthermore, some reviews about different organizational units exist in the current research. There are special problem areas in intensive care [34], emergency departments[35, 36] or in specialized treatment areas like radiation therapy [37]

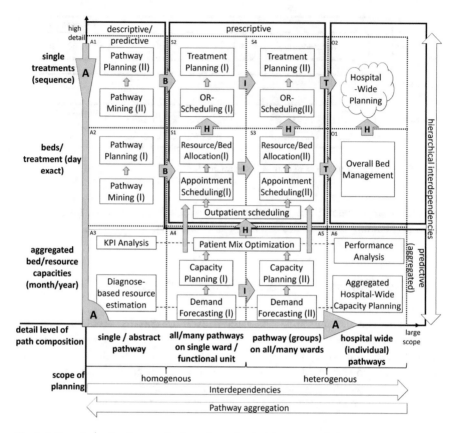

Fig. 1.6 Two-dimensional categorization scheme for decision support in hospitals

or outpatient scheduling in general. It is obvious that there are several review articles. Nevertheless all presented review articles share the same problem. Based on the focus of the review, there is only a categorization based on problem areas or organizational units.

Figure 1.6 shows the developed categorization scheme. Based on the previously presented abstraction levels and problem areas, the categorization scheme is implemented. On the y-axis, the abstraction levels of resources are shown, which represents the supply for any decision problem. The x-axis shows the different aggregation levels of patient pathways as demand side of the decision problem. Thus the scope of planning and basically the type of demand for resource capacity are located on this axis. By investigating the scheme, some general remarks about decision problems and interdependencies in a hospital environment are possible. In general with an increasing path aggregation the influence of interdependencies decreases. Nevertheless these interdependencies do not disappear. They are encoded into the aggregated pathways by making assumptions. Furthermore an increasing pathway aggregation increases the homogeneity of pathways and patient groups, which means that the corresponding decision problems do not have to deal with all possible interdependencies. In the opposite direction an increasing heterogeneity can be assumed, which leads to an increasing influence of interdependencies.

The presented scheme makes it possible to combine different types and levels of planning. In principle, planning areas that are oriented toward operational planning can be linked to strategic planning areas. In particular, the possible levels of detail can be described. In Fig. 1.6 it can be seen that quadrants S1, S2, S3, and S4 in the so-called scheduling sector are more oriented toward operational planning, while quadrants A1, A2, A3, A4, A5, and A6 in the analytic sector are more focused on strategic planning. Furthermore quadrants O1 and O2 in the overall planning sector deal with hospital-wide overall planning tasks.

In Fig. 1.6 it is obvious that resources and their considered granularity have a great influence on the specificity of problem areas. With a rising abstraction level of resources, a rising complexity of decision problems is observed. Furthermore a rising complexity results in a more strategical decision problem. While a more abstract view on resources results in more strategical problems, a changing abstraction level of patient paths only has a minor influence on the decision problem itself. However a more detailed view of patient pathways results in a more detailed planning problem inside a problem area.

A possible use of this classification scheme is the identification of subsequent decision tasks and the inclusion of interdependencies and flexibility. This will be shown using the example of the tasks in the area of arrow A in Fig. 1.6. The starting point for a planning consideration of the processes of a hospital is the planning of patient paths on an abstract level. In general, this involves planning the sequence of treatment steps to each diagnosis type. This can be done in two different ways (or in a combination). On the one hand, planning can be carried out by medical experts. Here, expertise and experience are integrated, which can lead to improvements in the treatment process of specific diagnosis types. On the other hand, planning can be based on historical data. Here a more descriptive planning takes place with the help of techniques of data/process mining. However, only the current actual state of the pathways can be mapped. A detailed discussion about pathway mining is done in [5]. By combining these two variants, computer-assisted planning of abstract pathways for different diagnosis can be done.

On the basis of these paths, an initial estimate of the required resources can be made. This is the general basis for capacity planning in a hospital, which can be done along the abstraction levels of the patient pathways in different areas, which are characterized by an increasing scope planning. As a result, planning is increasingly being driven toward hospital-wide planning. In general, it can be stated in advance that a simple summation of individual capacity plans at one level does not necessarily lead to planning at a higher level. Depending on the transition level, different interdependencies and additional constraints have to be considered.

In the transition from *"diagnosis-based bed/resource estimation"* to *"capacity planning (I) (CPI)"* it is not possible to sum up the results of the estimates for the individual diagnoses and to obtain a planning or estimation for the entire ward. In the aggregation, an additional intersection by patient-based interdependencies, i.e., the inclusion of gender segregation and the presence of isolation patients, must be reckoned with. Thus, the capacity requirement can be underestimated in a simple aggregation and problems can arise for operational use.

In the transition from *"capacity planning (I)"* to *"capacity planning (II) (CPII)"* further interdependencies have to be considered. In principle, planning in CPI takes place without considering shared resources, which are often planned as fixed resources for a ward. Thus each ward plans the capacity for shared resources separately. This leads to an overestimation of the required capacities. Simplified this can be described by the following formula: $\sum CPI + \alpha > CPII$. A separated planning causes unused resources at wards, which can be used better by integrating flexibility in an interdepartmental planning of (shared) resources. Generally the impact of shared resources is twofold. On the one hand there are passively shared resources like operating rooms or CT-scans. An isolated planning is too optimistic in respect for shared resources. The capacity of passively shared resources can be denoted by α. On the other hand there are actively shared resources like bed pools or staff pools. This kind of resources has to be planned in a flexible way across several wards in order to use their full potential. This flexibility strategies or the so-called flexibility cuts (ϕ) can have a huge impact on capacity planning. To enable a practical transition from CPI to CPII, we introduce the following relation: $\phi \cdot \sum CPI = CPII$ with $\phi < 1$. This makes it possible to expand the focus of planning without overestimating or underestimating the demand. A further advantage of this broader view of planning is the simple expandability to a hospital-wide focus. The transition to hospital-wide planning is not possible from CPI. However, it is much easier to plan hospital-wide from CPII. Since the most important interdependencies are already integrated in CPII, the expansion is done by considering organizational strategies. This results in aggregated predictive planning.

Basically sector A describes a descriptive or predictive approach for strategical hospital planning tasks. These tasks present an input for sector S, which describes a prescriptive view of planning tasks. This input is shown as arrow B in Fig. 1.6. For a hospital this is a crucial connection between strategical and tactical/operational planning areas. A high-quality planning in sector A is a mandatory requirement for the high quality of planning results in sector S. This can be discussed with the example of pathway mining and planning and its implication for sector S. There is a differentiation into three different variants. The first variant is a direct mining approach in A1. So the focus lies on a possible sequence of treatments. A second variant is the aggregation of treatments on a daily level. So not only the sequence is taken into account. Additionally the planned day of treatment inside a patient pathway is important for this mining and planning approach (quadrant A2). A third variant is a new combination of the previous discussed variants proposed by Helbig et al. [5]. This variant aims at the utilization of pathway mining and planning for the more operational scheduling sector. In order to avoid a too narrow set of input parameters for an operational planning, the exact location of each treatment inside a sequence is not important. Furthermore the exact day of treatment is not important either. In order to provide decent input for planning within this variant, only mandatory relations between treatments and a range of possible treatment days are considered. So this variant provides an appropriate input for other planning tasks without excluding possible solutions by providing too strict input parameters.

Furthermore it is interesting that the proposed scheme is able to link general planning tasks of a hospital with classical business analytics taxonomies. In general sector A engages, depending on the abstraction level, in a descriptive and a predictive analysis. A descriptive analysis deals with the question what happened while a predictive analysis deals with the more strategical question of what will happen. Planning tasks inside sector S can be seen as prescriptive planning of operations with strategic influences and can answer the question what should a hospital do.

1.4.2   Anticipation Issues of Decision Support Problems Within the Two-Dimensional Scheme

In order to emphasize the importance of the categorization scheme, we will present the differences of apparently similar planning problems in varying abstraction levels. First of all the differences between OR-scheduling (I) and OR-scheduling (II) is discussed. I describes the classical operating room scheduling. At this stage only the actual schedule is considered. In general different surgeries are scheduled based on their needed time and their importance. Based on the complexity of the considered problem, several other constraints like physicians' working hours or preparation times are implemented. Unfortunately a consideration of up- and downstream facilities is neglected during this decision task. However, for an operating room schedule suitable for operational use, a consideration of capacities of related units is crucial. By considering related units, a transition between I and II is performed. So the impact of interdependencies between the operating room and the related wards increases. A feasible schedule respects the capacity restrictions on other wards, and the complexity of the planning problem increases significantly but the possibility to transfer the schedule into operational work is increased. Another example is the problem area of appointment scheduling divided into appointment scheduling I and appointment scheduling II. Appointment scheduling (I) describes the task of assigning dates of admission to patients. Here it is important to assign all patients at one ward. Because of the isolated consideration of one ward, an optimal schedule could worsen the performance of other wards and corresponding organizational units. Especially if we assume multiple wards for several patients. Nevertheless this view leads to a demarcated and less complex decision problem regarding interdependencies. In order to decrease the impact of interdependencies between different wards or organizational units, an integrated consideration of multiple units is possible (appointment scheduling II). A scheduling for multiple wards is performed in one step. So a decision at one ward is automatically considered at another unit. So a trade-off between different solutions is found in order to achieve a global optimum. Hence it should be noted that advantages in an integrated planning lead to problems regarding complexity. So a detailed contemplation of interdependencies should be integrated in order to gain solutions that can be transferred to an operational use.

In general the transition from I to II independently from a problem area leads to a better modelling of interdependencies. So for most of the problem areas the equation $II = I + modelling\ of\ Interdependencies$ holds. Because of its transfer to a larger scope in direction of a global integrated planning (Quadrant O2), it is easier to reach the goal of an integrated planning platform.

Another relationship between planning areas is shown through arrow H in Fig. 1.6. Here a possible hierarchical relationship is described. Furthermore this relationship is described in Fig. 1.5 for the example of appointment scheduling and treatment planning. In a first step a daily planning of beds is performed as an input for a treatment planning. Additionally this treatment planning provides feedback for the bed management. So bed managers are able to adjust their planning based on this relationship. Another interesting hierarchical relationship is located between quadrant A4/A5 and S1/S3. Inside quadrant A4/A5 a connection between demand forecasting, capacity planning, and patient mix optimization is obvious. Based on a good and reliable demand forecast, a good capacity planning can be done. These two planning areas act as an input either for a strategical patient mix optimization or for a operational appointment scheduling and resource allocation. Furthermore patient mix optimization provides a corridor for a possible operational planning. An additional and mostly neglected possibility is the providing of a feedback loop between sector S and quadrants A4/A5. With the help of anticipating techniques for operational planning, a readjustment of strategical decisions based on anticipated operational decisions is possible.

Arrow T in the categorization scheme describes the transition to a hospital-wide planning view. This basically means the enhancement of department-wide planning approaches with organizational characteristics and strategies.

An overall planning view is mainly influenced by overall organizational aspects, for example, organizing pools of resources or managing rearrangements of functional units. Furthermore this includes a possible adjustment of concepts or ideas to changing conditions of a hospital, which can be done in two different ways. First the changes or reorganizations can be done permanently in order to cope with strategical aspects. The second way is the rearrangement in case of emergency situations. For example, if an epidemic outbreak occurs, the hospital management is able to use organizational strategies, which promotes flexibility on the top planning level, which can be an ad hoc rearrangement of nursing staff and physicians or a reallocation of resources.

1.4.3 Classifying Problem Areas Based on the Categorization Scheme

For a more detailed categorization, a further literature analysis based on existing reviews and newer additional papers is done to achieve two main goals. First of all we like to achieve a more detailed classification of problem areas. As a second goal an identification of currently unregarded problem areas in operations research is desired.

From [28] it is possible to extract general problem areas for operations research in healthcare (see Fig. 1.7). In general problem areas can be divided into two subsections. First we consider all planning problems. These problems consist of issues about demand forecasting, location planning, capacity planning, and treatment planning. In a second area healthcare management and logistic problems are considered. From an operational point of view, scheduling is an important topic, in detail patient or resource scheduling. Patient scheduling can be divided into inpatient and outpatient scheduling. For resource scheduling a more detailed partition is needed. Hence there is a split into nurse, physician, operating room, and other scheduling problems. For a further classification insights of the review articles and other recent research literature, a two-dimensional division of problem areas based on abstraction levels is needed. The body of literature is scanned based on the described abstraction levels, and their problem areas are included into the two-dimensional categorization scheme. So a relevant map of problem areas is built and described in the next paragraph.

Within these analyses, an evaluation of the current direction of research is possible. Table 1.2 shows a literature review of current scientific contributions mostly published after 2017. For selected problem areas in different quadrants of the categorization scheme, a distinction between optimization- and simulation-based contributions is done. This review is used to point up current research directions in selected problem areas in order to gain general insights. While using

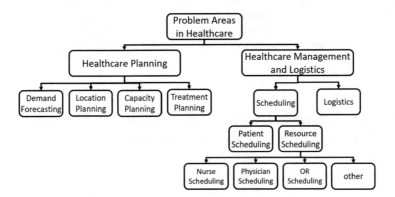

Fig. 1.7 General problem areas in health care

Table 1.2 Assignment from the literature to quadrants and problem areas

Quadrant	Problem area	Optimization	Simulation
S1	Outpatient scheduling	[39–44]	[45, 46]
S1	Appointment scheduling	[13, 49]	[13]
S1	Resource allocation	[50–54]	–
S3	Bed allocation	[55]	
S3	Patient scheduling		[56]
S4	OR-scheduling	[38]	–
A3	Diagnosis-based resource estimation	[47, 48]	–
A3	KPI analysis	–	[59, 62–65]
A4	Capacity planning	[58]	[59–61]
A4 ∣ A5	Patient mix optimization	–	[66, 67]
A6	Performance analysis	–	[57, 68–70]
A6	Capacity planning	–	[71]

the presented categorization scheme in combination with Table 1.2, it is obvious that current research is focused on the quadrants located in the left and center of the categorization scheme. Because of that concentration, it is noticeable that in most of the current research interdependencies are left out. Mostly research about single organizational units is done. So a neglection of interdependencies results in a lack of practicality of the investigated decision problems. Furthermore the usage of abstract pathways, which are not fully able to reproduce the behavior of patients, decreases the practicality. So current research activities can be seen as feasibility studies where the last step to a practical implementation is missing because of an insufficient consideration of interdependencies.

Through Table 1.2 another important insight is given. Optimization techniques are mostly used for rather tactical or operational problem areas with more abstract patient paths. For more complex or strategical problem areas, simulation techniques are used. Nevertheless the most important insight from a resource point of view, as stated before, is the concentration on abstract paths and single organizational units. Furthermore most of the scanned literature is focused on single decision problems. Hierarchical interdependencies are mostly left out. In order to discuss a possible usage of interdependencies, two kinds of case studies will be presented in the next section.

1.5   Case Studies in Hospital Decision Support: Flexibility Strategies Coping with Interdependencies

After presenting the categorization scheme and classifying the current research literature, two case studies are presented below, which exploit the positive aspects of interdependencies at different levels of abstraction.

1.5.1 Simulation Case Study

Our research team developed a valid simulation system based on data analysis of one-year data of a university hospital to increase flexibility in bed management [2]. Because of the introduction of the German Diagnosis-Related Group (G-DRG) system, some major changes for hospitals occurred in Germany. Based on these groups, the billing system of hospitals changed significantly and the economic pressure rose dramatically. Furthermore hospitals were forced to collect and transmit standardized case data. This data is used to improve the G-DRG system. This data is fundamental for further analysis of clinical pathways and performance of hospitals. CP forms sequences of ward stays and treatments to be performed during a patient's hospitalization under consideration of all relevant resources like beds, staff, and moveable medical equipment. Through this analysis, it is possible to find rules for patient arrivals, length of stay, and internal transfers of patients and to calculate empirical distributions based on these rules.

Based on these findings, we found that simple, yet sufficiently detailed, pathways are constructed in order to build a hospital-wide simulation model (see Fig 1.8). It is notable that the presented paths are abstract pathways. While using the simulation model, these abstract paths transform into individual patient pathways. For each patient inside the simulation model, individual transfer probabilities are calculated from distributions in order to construct patient paths. Furthermore the simulation model uses the whole hospital infrastructure as an input. The core calculation component of the presented case study performs a daily bed allocation. Here a heuristic is used in order to distribute patients to possible room, respectively, beds. While calculating the allocation, several constraints based on capacity-based interdependencies were taken into account, namely, the consideration of gender separation and the isolation of patients with multiresistant germs.

With the help of simulation techniques, the problem area is expanded. Now it is possible to make strategical capacity decision or to evaluate these decision based on operational decision-making. With the expansion of the focus, more interdependencies were taken into account. Here process- and capacity-based interdependencies between organizational units are considered. This is a concept, which distinguishes the presented model from other simulation-based approaches

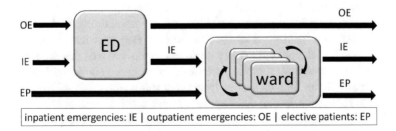

Fig. 1.8 Clinical pathway for simulation

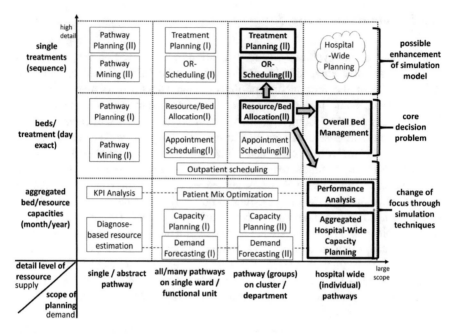

Fig. 1.9 Classification of case studies

in the state of the art. So we are able to tackle different kind of problem areas on different abstraction levels within the same model. Furthermore it is possible to enhance the model by integrating components for different kind of treatment scheduling areas. Logically this is the next step in development because of the hierarchical interdependencies between bed allocation and treatment scheduling. The categorization of the presented case study and its implication is shown in Fig. 1.9.

One important milestone of this case study is the validation of the model for hospital-wide data over one year. We notice that empirical distributions generated from standardized data are accurate enough to model reality in a strategical context. A detailed discussion about validating this model is laid out in [3]. With the help of a validated simulation model, we are able to investigate several ideas in order to promote flexibility in hospital ward and bed management.

A first approach to improve flexibility was the opportunity to lend idle bed capacities between nearby wards (in this primary step without consideration of treatment needs of the patient). A key performance indicator of this scenario is the average waiting time until a patient is transferred to his or her desired ward. With the opportunity to use idle capacities from other wards, the waiting time decreased up to 10 h. However a feared disadvantage for some wards did not occur. Another possible way to improve flexibility was to pool capacities from intensive care and intermediate care wards. The findings are similar to the first scenario. The waiting time for transfers decreased significantly. Furthermore, occupancy rates of some

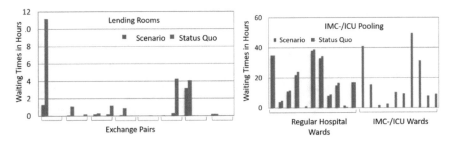

Fig. 1.10 Experimental results of case study 1

hospital wards are rising, while other wards with a high initial occupancy rate notice a decrease in their utilization. Additionally the presented model shows a decrease in occupancy bottlenecks because of a more flexible use of resources. This was a first step toward new thoughts about flexibility in ward and bed management (see Fig. 1.10).

Further discussions with the management of a university hospital led to the concept of ward occupancy cluster in hospitals in order to incorporate medical issues in the flexibility approach. In general, this introduces a new improved version of the opportunity to lend idle capacities with respect to organizational and treatment-based restrictions. In [3], a generic simulation-based decision support application in Simio is developed that enables to evaluate flexible ward clusters in hospital occupancy management. Basically an occupancy cluster is a union of wards that nurses similar patients. Each cluster will contain a number of wards. Within a cluster, all kinds of patients could be nursed by each containing ward. For example, a surgery cluster contains all hospital wards dealing with surgeries (e.g., heart surgery, thoracic surgery, and plastic surgery). At first, every ward tries to use its own beds. If there is no more capacity for another patient, this patient will be admitted to another ward of the cluster that has available capacities. In theory this approach should be able to distribute utilization peaks evenly between wards.

Two major key performance indicators are investigated in simulation studies in order to evaluate this strategy. In general we analyze the utilization rate and the number of waiting or rejected patients. The evaluation of this strategy shows that occupancy clusters are able to distribute occupancy peaks (see Fig. 1.11). The peak utilization of the wards did not rise above 100%. Moreover the number of waiting or rejected patients decreases significantly. The performance of this indicator depends on the evaluated cluster configuration. However, it should be noted that every cluster configuration studied significantly improves the status quo. This is a clear indication that even limited flexibility is better than no flexibility at all. In future research the location of the wards can be included in this approach in order to smoothen the utilization of other resource like staff and moveable medical tools.

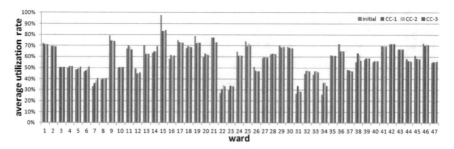

Fig. 1.11 Experimental results of case study 2

1.5.2 Optimization Case Study

Previous case studies aimed at an improvement of operations under flexibility from simulation perspective and a capacity-related point of view. Another way to improve flexibility is to optimize the treatment of a patient based on daily planning of treatments [4]. This study proposes a hierarchical integer linear programming (ILP)-based approach for the tactical day-level scheduling of clinical pathways. The approach uses a multicriteria objective function considering several patient- and hospital-related aspects and allowing to employ our approach in various scenarios. Furthermore, compared to other approaches, it considers several compliance discussed in Sect. 1.1. Basically this case study is located in the third and 7th quadrant of the categorization scheme. Because of its hierarchical architecture, this approach is located in several quadrants. On the top level a daily resource allocation is done, while on the second layer a detailed scheduling of treatments is done. So this case study is considered as a hybrid approach with respect to the presented categorization scheme (see Fig. 1.12).

This approach is evaluated with real-world data from a German university hospital showing that our approach is able to solve instances with a planning horizon of one month exhibiting 1088 treatments and 302 ward stays of 286 patients. These experiments show that our approach is capable of obtaining high-quality solutions within a reasonable amount of time for a tactical planning problem. Furthermore, the detailed evaluation shows that our model is capable of considering different scenarios such as achieving a smooth allocation of resources over the planning horizon and establishing a high resource allocation in the beginning of the horizon.

In the first scenario, the hospital aims at establishing a smooth resource allocation throughout the planning horizon. A smooth resource allocation establishes a similar level of flexibility to react to emergency cases on each day of the planning horizon. In the second scenario, the hospital aims at establishing a high resource utilization at the beginning of the planning horizon. On the one hand, shifting workload into the near future provides more scheduling flexibility for the more distant future to cope with uncertainties, e.g., due to complications and emergencies; on the other hand, this flexibility enables the hospital to seize opportunities for admitting a higher number of elective patients in the future.

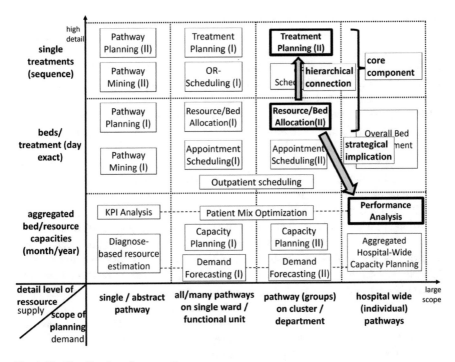

Fig. 1.12 Classification of case studies

1.6 Flexibility Strategies in Hospital-Wide Bed Management

In the previous section two case studies using flexibility strategies for resource usage was described. In order to discuss the strategies in detail, we first derive a definition of flexibility and transfer it to bed management. Afterward some current strategies from the case studies are discussed, and further research possibilities are described.

1.6.1 Definition of Flexibility in the Context of Bed Management

Flexibility has manifold definitions. Depending on the field of research or interest, the definition differs significantly. Because of the similarity between production companies and hospitals, we are able to transfer and enlarge the definition of flexibility from a production point of view to a hospital environment. Flexibility in general describes the ability of a system to change or react with little penalty in time, effort, cost, or performance to a varying environment [72]. This is a rather abstract definition of flexibility. By considering the resource point of view, flexibility can be located on the supply side of operations [73]. In order to support this point of view, product flexible manufacturing capacities are introduced in [74, 75]. That means different products can be produced on the same machine capacity with little to no extra effort except costs for investments in flexible capacities. One of the key areas of flexibility in a production environment is the possible response to uncertainty. Organizations need to cultivate innovative capacities to promote higher flexibility. Integrating flexible and fixed structures can lead to lower costs and increased performance in operations [76]. By expanding the focus on supply chain flexibility, [76] argues that flexibility is repositioned from an intra- to an interorganizational concept. In general flexibility in manufacturing organizations is a multidimensional process across various functional layers of an organization. But flexibility in and of itself is not a solution. Instead it is a way to achieve a specific goal by delivering a specific response to an environment.

This definition of flexibility can be transferred to flexibility in a hospital environment. Especially attributes of flexible manufacturing capacities and supply chain flexibility can be transferred to hospital bed management. Patients can be seen as products, and in this case beds are considered as product flexible manufacturing capacities. Additionally if a hospital-wide bed management is regarded, principles of supply chain flexibility presented in the literature can be used in order to improve the performance of the bed management. Unfortunately it is not possible to transfer mechanisms of production flexibility directly. Because of the greatly enlarged patient uncertainty, the methods have to be expanded. In the next subsection current flexibility strategies in hospital bed management will be discussed.

1.6.2   Current State of Flexibility

In current decision support practice in hospitals, a low grade of flexibility is observed. In general each problem area, respectively, each organizational unit is treated separately. So possible advantages from interdependencies are neglected. Within this chapter, we will focus on occupancy-based flexibility strategies in order to promote a more efficient way of using existing resources because they are rather similar to the above-presented definition of flexibility. A general overview about different occupancy-based strategies is shown in Fig. 1.13, which shows the classification according to an increasing flexibility research line. By assuming general

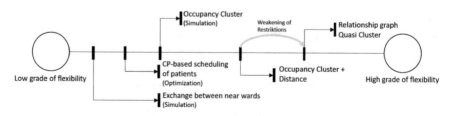

Fig. 1.13 Classification of present and future methods in a flexibility context

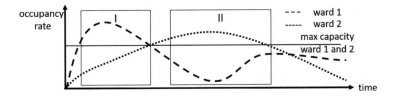

Fig. 1.14 Application areas of occupancy-based flexibility strategies

equality between beds at different wards, it is possible to distribute occupancy peaks from one ward to another. In Fig. 1.14 some occupancy rate developments over time are shown. In I the maximum possible capacity of ward one is exceeded and for ward two capacity is left unused. Practically this leads to rejection of patients or waiting time at ward one and unused following resources like physicians and nurses at ward two. In II it is vice versa.

The first flexibility strategy tries to extract advantages from this occupancy development. At the first stage of flexibility, the simple exchange of empty beds between near wards is used. In I ward one is not able to admit every patient. So this ward uses empty beds of ward two in order to ensure an admission of every patient. So with little to no effort a simple way of increasing the usage of resources is introduced. The downfall of this strategy is the limited amount of flexible resource capacity. Only a few beds of each ward can be used as backup for other wards. Additionally the number of lending partner is limited because of the precondition of a neighborhood of wards. Another flexibility strategy is the pooling of capacities, for example, from intensive care and intermediate care wards. First experimental results of both strategies imply a significant decrease of waiting times and the occurrence of occupancy bottlenecks without the appearance of disadvantages for some wards.

Encouraged by these findings and by further discussion with medical management experts, occupancy clusters are introduced in order to promote flexibility. Hereby the similarity of wards is considered while developing possible cluster configuration. The possibility of using beds from other wards is increased because potentially every bed can be used for other wards. Our study results regarding occupancy-based flexibility suggests a better resource usage of existing resources without investing in new beds. Occupancy clusters are able to distribute occupancy peaks evenly and moreover are able to decrease the number of waiting or rejected patients significantly. Nevertheless the performance of a cluster configuration strongly depends on its compilation. Different occupancy rates at the wards have a great impact on the possibility of using additional beds. If occupancy rates like in Fig. 1.14 are realized, it is simple to use beds from other wards. If the development of occupancy rates are too similar between wards, it is possible that no additional capacity is available when needed because these wards have their occupancy peaks at the same time. A great advantage of the cluster strategy is the possibility of combining multiple wards because the negative impact of too similar occupancy rates can be compensated through a rising number of combined wards. Based on these promising results and in order to increase flexibility, more advanced strategies in further research will be discussed in the following section.

1.6.3   Further Research

As noted before we could combine the ward occupancy cluster approach with a consideration of locations of the hospital wards. This is the first step in order to improve the performance of our approach. Another possible way to improve performance of such a cluster configuration is to weaken the organizational restrictions. Several given cluster configurations from the hospital management imply that there are no strict medical reasons for a special cluster configuration or for a strict border between occupancy cluster. From a research point of view, it is possible to determine a possible softening of organizational restrictions in several ways. Regarding different cluster configuration, one possibility is the determination of intersection set of occupancy cluster, the so-called core cluster (Clique). By investigating further connections inside a cluster configuration, it is possible to calculate additional connections between core clusters.

A second possibility to weaken strict, but not necessary, organizational restrictions is calculating the similarity or dissimilarity of hospital wards based on their range of diagnosis and treatments. First results show that we are able to quantify the pairwise similarity between wards and to select a threshold of similarity. Based on that a similarity matrix for hospital wards can be calculated.

Both approaches (intersection of clusters and similarity matrix) allow us to break out of our classical cluster structure with strict delimitation. Based on those findings, we are able to implement a relationship graph in order to support the resource allocation process. Additionally this approach supports soft delimitation based on the hospital situation. For example, if there is an epidemic or a huge accident, the cluster delimitation (or the so-called softening degree) can be increased accordingly. Regarding this, a quasi-cluster structure can be updated for hospital-wide emergency situations. Figure 1.15 shows the resulting relationship graph of a cluster configuration with breakout relations. If one cluster is fully occupied, we are then able to use capacities of other clusters by utilizing these breakout relations. Graph-theoretically, we have a dependency graph of cliques with few intercluster connections. With the help of this concept, flexibility regarding hospital ward and bed management can be significantly improved without fully neglecting organizational restrictions and acceptancy of ward medical staff. Besides this, a further flexibility dimension can be added to our concept: If there is a case of an epidemic spread or of a huge accident, the cluster delimitation can be further softened in order to treat higher patient rates. Thus, for hospital-wide emergency situations, the cluster structure can be easily updated using precalculated ward similarities by increasing softening degrees resulting in temporarily more connected core clusters.

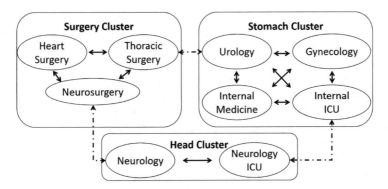

Fig. 1.15 Relationship graph for a hospital with quasi-cluster structure

1.7 Conclusion

Within this chapter, we reached our desired goals. For the first time, a detailed consideration of interdependencies inside a hospital is done. Furthermore these interdependencies are discussed with respect to their problem area and their possible impact on decision problems. With the help of new abstraction levels of patient pathways and resource, a classification of problem areas and current research is done. Here a concentration on certain abstraction levels is obvious. Optimization is mainly used with abstract patient pathways and with single organizational units. Hence optimization is mostly used for tactical or operational problem areas. The analysis of the current research literature shows that more complex paths are mainly tackled by simulation. In some research papers there is a consideration of multiple organizational units. But in general it is notable that most decision problems are discussed based on single organizational units or abstract pathways. That is why it is difficult to transport these studies from research to practice. In order to ensure a practical use, applications for decision support have to respect as much interdependencies as possible.

With the presentation of several case studies, possible ways of using inter-dependencies to promote flexibility are shown. In the simulation case studies, interdependencies of patient pathways are used to improve flexibility for the resource allocation process. Within the optimization case study, a possible imple-mentation of flexibility inside the treatment planning process is presented. Here process interdependencies are used to improve the treatment planning.

Nevertheless it is vital for the operational use of current research in decision support in hospitals that researcher understand the influence of interdependencies and that more complex decision models are implemented with respect to more organizational units and more complex patient paths. This could lead to a general improvement of practical usage of current approaches in the operations research community.

Another important aspect is that optimization is mostly used for operational problem areas. In contrast, mainly simulation is used for more strategical questions. Unfortunately in nearly every simulation study the operational decision-making is neglected or is done in a greatly simplified way. Because of the rising integration of decision support systems and methods into the operational work of a hospital, this kind of decision-making should be respected in strategical considerations. Hence the operations research and especially the simulation community should move to a simulation of the implemented decision support methods inside a simulation study by using the same components as in operational use. Because of that an intelligent anticipating planning behavior can be simulated. Hence a practical implementation of study results is simplified. Moreover local maxima should not lead to a worsen of the overall performance of a hospital. So an organization-wide simulation with integrated decision support systems can lead to a better understanding of which interdependencies are important for a hospital setting. It is possible to decide which interdependencies should be included into the optimization and which interdependencies could be anticipated. In a next research step we develop a hierarchical anticipating approach for integrated admission scheduling and resource allocation in a hospital.

References

1. Mellouli, T., Stoeck, T.: Synergies between predictive mining and prescriptive planning of complex patient pathways considering process discrepancies for effective hospital wide decision support. In:Masmoudi, M., Jarboui, B., Siarry, P. (eds.) Artificial intelligence and Data mining in healthcare, SPRINGER (2020)
2. Helbig [Schwarz], K., Mellouli, T., Stoeck, T., Gragert, M., Jahn, P.: Simulation stationsübergreifender Patientenflüsse zur Evaluation flexibler Bettenbelegungsszenarien aufgrund der Jahresdatenanalyse eines Universitätsklinikums. In: MKWI 2014 – Multikonferenz der Wirtschaftsinformatik: 26. – 28. Februar 2014 in Paderborn: Tagungsband, 749–762. University of Paderborn (2014)
3. Helbig [Schwarz], K., Stoeck, T., Mellouli, T.: A Generic Simulation-Based DSS for Evaluating Flexible Ward Clusters in Hospital Occupancy Management. In: IEEE (eds.) Proceedings of the 48th Annual Hawaii International Conference on System Sciences, pp. 2923–2932 (2015)
4. Schwarz, K., Römer M., Mellouli T.: A Data-Driven Hierarchical MILP Approach for Scheduling Clinical Pathways: A Real-World Study from a German University Hospital To appear in BUSINES RESEARCH (2016)
5. Helbig [Schwarz], K., Römer, M., Mellouli. T.: A Clinical Pathway Mining Approach to Enable Scheduling of Hospital Relocations and Treatment Services. In Business Process Management, ed. Hamid Reza Motahari-Nezhad, Jan Recker, and Matthias Weidlich, 9253, pp242–250. Cham: Springer International Publishing (2015)
6. Stoeck, T., Mellouli, T., Schwarz, K.: Case Studies in hospital ward and bed management: decision support by simulation and optimization under flexibility. AFROS 2018
7. Gluchowski, P., Gabriel, R., Dittmar, C.:Management Support Systeme und Business Intelligence. Springer, Berlin, Heidelberg (2008)
8. Helbig [Schwarz], K.: Ein datengetriebenes System auf Basis klinischer Pfade zur Entscheidungsunterstützung für Ressourcenplanung in Krankenhäusern - Prozess-Mining, Optimierung und Simulation (2016), http://dx.doi.org/10.25673/1922

9. Schumann, C., Schieber, A., Hilbert, A.: Moderne Entscheidungsunterstützung im Kranken-haus – Business Intelligence meets Healthcare. HMD Praxis der Wirtschaftsinformatik **53/3**, 287–297 (2016)
10. van Merode, G. G., Groothuis, S., Hasman, A.: Enterprise resource planning for hospitals. International Journal of Medical Informatics **73/6**, 493–501 (2004)
11. Abraham, G., Byrnes, G. B., Bain, C. A.: Short-Term Forecasting of Emergency Inpatient Flow. IEEE Transactions on Information Technology in Biomedicine **13/3**, 380–388 (2009)
12. Atkinson, J., Wells, R., Page, A., Dominello, A., Haines, M., Wilson, A.: Applications of system dynamics modelling to support health policy. Public Health Research & Practice **25/3**, (2015)
13. Bakker, M., Tsui, K.: Dynamic resource allocation for efficient patient scheduling: A data-driven approach. Journal of Systems Science and Systems Engineering **26/4**, 448–462 (2017)
14. Green, L. V.: Capacity Planning and Management in Hospitals. In: Brandeau, M. L., Sainfort, F., Pierskalla, W. P. (eds.) Operations Research and Health Care, pp. 15–41. Kluwer Academic Publishers, Boston (2005)
15. Kolker, A.: Interdependency of Hospital Departments and Hospital – Wide Patient Flows. In: Hall, R. (eds.) Patient Flow, pp. 43–63. Springer, Boston (2013)
16. Lamothe, L., Dufour, Y.: Systems of interdependency and core orchestrating themes at health care unit level: A configurational approach. Public Management Review **9/1**, 67–85 (2007)
17. Burke, E. K., Curtois, T., Qu, R., Vanden Berghe, G.: A scatter search methodology for the nurse rostering problem. Journal of the Operational Research Society **61/11**, 1667–1679 (2010)
18. Burke, E. K., De Causmaecker, P, Berghe, G. V., Van Landeghem, H.: The State of the Art of Nurse Rostering. Journal of Scheduling **7/6**, 441–449 (2004)
19. Roche, K. T., Rivera, D. E., Cochran, J. K.: A control engineering framework for managing whole hospital occupancy. Mathematical and Computer Modelling **55/3-4**, 1401–1417 (2012)
20. Bartz, M.: Patientenpfade Ein Instrument zur Prozessoptimierung im Krankenhaus. VDM Verlag Dr. Müller (2006)
21. Müller, H., Schmid, K., Conen, D.: Qualitätsmanagement: Interne Leitlinien und Patien-tenpfade. Medizinische Klinik **96/11**, 692–697 (2001)
22. Holler, T., Conen, D.: Kostenbasierte Behandlungspfade. In: Albrecht, D. M., Töpfer, A. (eds.) Erfolgreiches Changemanagement im Krankenhaus, pp. 167–179. Springer, Berlin Heidelberg (2006)
23. Holler, T.,Conen, D.: Prozessorientierung – Analyse und Optimierung von Wertschöp-fungsprozessen. In: Albrecht, D. M., Töpfer, A. (eds.) Handbuch Changemanagement im Krankenhaus, pp. 217–228, Springer, Berlin Heidelberg (2017)
24. Wicke, C., Teichmann, R., Holler, T., Rehder, F., Becker, H. D.: Entwicklung und Einsatz von Patientenpfaden in der Allgemeinchirurgie. Der Chirurg **75/9**, 907–915 (2004)
25. Helbig, K.: Zeitplanung für Patientenpfade unter Berücksichtigung von Betten-,Behandlungskapazitäten und Fairnisskriterien. In: Alt, R., Eisenecker, U., Franczyk, B., Heyden, K. (eds.) Forschungsberichte des Instituts für Wirtschaftsinformatik der Universität Leipzig, pp. 34–44. Leipzig (2011)
26. Proudlove, N. C.: Can good bed management solve the overcrowding in accident and emergency departments?. Emergency Medicine Journal **20/2**, 149–155 (2003)
27. Black, D.: Average length of stay, delayed discharge, and hospital congestion. BMJ **325/7365**, 610–611 (2002)
28. Rais, A.,Viana, A.: Operations Rese3arch in Healthcare: a survey. International Transactions in Operational Research **18/1**, 1–31 (2011)
29. Fone, D., Hollinghurst, S., Temple, M., Round, A., Lester, N., Weightman, A., Roberts, K., Coyle, E., Bevan, G., Palmer, S.: Systematic review of the use and value of computer simulation modelling in population health and health care delivery. Journal of Public Health **25/4**, 325–335 (2003)
30. Günal, M. M., Pidd, M.: Discrete event simulation for performance modelling in health care: a review of the literature. Journal of Simulation **4/1**, 42–51 (2010)

31. Cardoen, B., Demeulemeester, E., Beliën, J.: Operating room planning and scheduling: A literature review. European Journal of Operational Research **201/3**, 921–932 (2010)
32. Jack, E. P., Powers, T. L.: A review and synthesis of demand management, capacity management and performance in health-care services. International Journal of Management Reviews **11/2**, 149–174 (2009)
33. Baru, R. A., Cudney, E. A., Guardiola, I. G., Warner, D. L., Phillips, R. E.: Systematic Review of Operations Research and Simulation Methods for Bed Management. Proceedings of the 2015 Industrial and Systems Engineering Research Conference (2015)
34. Bai, J., Fügener, A., Schoenfelder, J., Brunner, J. O.: Operations research in intensive care unit management: a literature review. Health Care Management Science **21/1**, 1–24 (2018)
35. Saghafian, S., Austin, G., Traub, S. J.: Operations research/management contributions to emergency department patient flow optimization: Review and research prospects. IIE Transactions on Healthcare Systems Engineering **5/2**, 101–123 (2015)
36. Gul, M., Guneri, A. F.: A comprehensive review of emergency department simulation applications for normal and disaster conditions. Computers & Industrial Engineering **83**, 327–344 (2015)
37. Vieira, B., Hans, E. W., van Vliet-Vroegindeweij, C., van de Kamer, J., van Harten, W.: Operations research for resource planning and -use in radiotherapy: a literature review. BMC Medical Informatics and Decision Making **16/1**, (2016)
38. Bai, M., Storer, R. H., Tonkay, G. L.: A sample gradient-based algorithm for a multiple-OR and PACU surgery scheduling problem. IISE Transactions **49/4**, 367–380 (2017)
39. Conforti, D., Guerriero, F., Guido, R.: Optimization models for radiotherapy patient scheduling. 4OR **6/3**, 263–278 (2008)
40. Diamant, A., Milner, J., Quereshy, F.: Dynamic Patient Scheduling for Multi-Appointment Health Care Programs. Production and Operations Management **27/1**, 58–79 (2018)
41. El-Sharo, M., Zheng, B., Yoon, S. W., Khasawneh, M. T.: An overbooking scheduling model for outpatient appointments in a multi-provider clinic, Operations Research for Health Care **6**, 1–10 (2015)
42. Maschler, J., Hackl, T., Riedler, M., Raidl, G. R.: An Enhanced Iterated Greedy Metaheuristic for the Particle Therapy Patient Scheduling Problem. MIC/MAEB (2017)
43. Maschler, J., Raidl, G. R.: Particle therapy patient scheduling with limited starting time variations of daily treatments. International Transactions in Operational Research **00**, 1–22 (2018)
44. Papi, M., Pontecorvi, L., Setola, R., Clemente, F.: Stochastic Dynamic Programming in Hospital Resource Optimization. In: Sforza, A., Sterle, C. (eds.) Optimization and Decision Science: Methodologies and Applications, pp. 139–147 Springer, (2017)
45. Gocgun, Y.: Simulation-based approximate policy iteration for dynamic patient scheduling for radiation therapy. Health Care Management Science **21/3**, 317–325 (2018)
46. Bruballa, E., Wong, A., Rexachs, D., Luque, E., Epelde, F.: Scheduling model for noncritical patients admission into a hospital emergency department. In: IEEE (eds.) 2017 Winter Simulation Conference (WSC), pp. 2917–2928 (2017)
47. Davis, S., Fard, N.: Theoretical bounds and approximation of the probability mass function of future hospital bed demand. Health Care Manag Sci (2018). https://doi.org/10.1007/s10729-018-9461-7
48. Luo, L., Luo, L., Zhang, X., He, X.: Hospital daily outpatient visits forecasting using a combinatorial model based on ARIMA and SES models. BMC Health Services Research **17/1**, – (2017)
49. Patrick, J., Puterman, M. L., Queyranne, M.: Dynamic Multipriority Patient Scheduling for a Diagnostic Resource. Operations Research **56/6**, 1507–1525 (2008)
50. Bolaji, A. L., Bamigbola, A. F., Shola, P. B.: Late acceptance hill climbing algorithm for solving patient admission scheduling problem. Knowledge-Based Systems **145**, 197–206 (2018)
51. Turhan, A. M., Bilgen, B.: Mixed integer programming based heuristics for the Patient Admission Scheduling problem. Computers & Operations Research **80**, 38–49 (2017)

52. Bastos, L. S. L., Marchesi, J. F., Hamacher, S., Fleck, J. L.: A mixed integer programming approach to the patient admission scheduling problem. European Journal of Operational Research (2018), https://doi.org/10.1016/j.ejor.2018.09.003
53. Ogulata, S. N., Koyuncu, M., Karakas, E.: Personnel and Patient Scheduling in the High Demanded Hospital Services: A Case Study in the Physiotherapy Service. Journal of Medical Systems **32/3**, 221–228 (2008)
54. Guido, R., Solina, V., Conforti, D.: Offline Patient Admission Scheduling Problems. In: Sforza, A., Sterle, C. (eds.) Optimization and Decision Science: Methodologies and Applications, pp. 129–137 Springer, (2017)
55. Aringhieri, R., Landa, P., Mancini, S.: A Hierarchical Multi-objective Optimisation Model for Bed Levelling and Patient Priority Maximisation. In: Sforza, A., Sterle, C. (eds.) Optimization and Decision Science: Methodologies and Applications, pp. 113–120 Springer, (2017)
56. Paulussen, T. O., Jennings, N. ., Decker, K. S., Heinzl, A.: Distributed Patient Scheduling in Hospitals. IJCAI'03 Proceedings of the 18th international joint conference on Artificial intelligence, 1224–1229 (2003)
57. Busby, C. R., Carter, M. W.: Data-driven generic discrete event simulation model of hospital patient flow considering surge. In: IEEE (eds.) 2017 Winter Simulation Conference (WSC), pp. 3006–3017 (2017)
58. Côté, M. J.: A note on Bed allocation techniques based on census data. Socio-Economic Planning Sciences **39/2**, 183–192 (2005)
59. van de Vrugt, N. M., Schneider, A. J., Zonderland, M. E., Stanford, D. A., Boucherie, R. J.: Operations Research for Occupancy Modeling at Hospital Wards and Its Integration into Practice. In: Kahraman, C., Topcu, Y. I. (eds.) Operations Research Applications in Health Care Management, pp. 101–137 Springer, (2018)
60. Carmen, R., Defraeye, M., Van Nieuwenhuyse, I.: A Decision Support System for Capacity Planning in Emergency Departments. International Journal of Simulation Modelling **14/2**, 299–312 (2015)
61. Cochran, J. K., Roche, K.: A queuing-based decision support methodology to estimate hospital inpatient bed demand. Journal of the Operational Research Society **59/11**, 1471–1482 (2008)
62. Kuo, Y.-H., Leung, J. M.Y., Graham, C. A., Tsoi, K. K.F., Meng, H. M.: Using simulation to assess the impacts of the adoption of a fast-track system for hospital emergency services. Journal of Advanced Mechanical Design, Systems, and Manufacturing **12/3**, – (2018)
63. Persson, M., Persson, Jan A.: Health economic modeling to support surgery management at a Swedish hospital. Omega **37/4**, 853–863 (2009)
64. Cayirli, T., Dursun, P., Gunes, E. D.: An integrated analysis of capacity allocation and patient scheduling in presence of seasonal walk-ins. Flex Serv Manuf J (2018). https://doi.org/10.1007/s10696-017-9304-8
65. Luo, L., Liu, C., Feng, L., Zhao, S., Gong, R.: A random forest and simulation approach for scheduling operation rooms: Elective surgery cancelation in a Chinese hospital urology department. The International Journal of Health Planning and Management **33/4**, 941–966 (2018)
66. Gedik, R., Zhang, S., Rainwater, C.: Strategic level proton therapy patient admission planning: a Markov decision process modeling approach. Health Care Management Science **20/2**, 286–302 (2017)
67. Freeman, N., Zhao, M., Melouk, S.: An iterative approach for case mix planning under uncertainty. Omega **76**, 160–173 (2018)
68. Landa, P., Sonnessa, M., Tànfani, E., Testi, A.: Multiobjective bed management considering emergency and elective patient flows. International Transactions in Operational Research **25/1**, 91–110 (2018)
69. Schneider, A. J. T., Besselink, P. L., Zonderland, M. E., Boucherie, R. J.: Allocating Emergency Beds Improves the Emergency Admission Flow. Interfaces **48/4**, 384–394 (2018)
70. Alvarado, M. M., Cotton, T. G., Ntaimo, L., Pérez, E., Carpentier, W. R.: Modeling and simulation of oncology clinic operations in discrete event system specification. Simulation **94/2**, 105–121 (2018)

71. Devapriya, P., Strömblad, C. T. B., Bailey, M. D., Frazier, S., Bulger, J., Kemberling, S. T., Wood, K. E.: StratBAM: A Discrete-Event Simulation Model to Support Strategic Hospital Bed Capacity Decisions. Journal of Medical Systems **39/10**, – (2018)
72. Upton, D. M.: The management of manufacturing flexibility. California Management Review **36/2**, 72–89 (1994)
73. Chod, J., Rudi, N.: Resource Flexibility with Responsive Pricing. Operations Research **53/3**, 532–548 (2005)
74. Fine, C. H., Freund, R. M.: Optimal Investment in Product-Flexible Manufacturing Capacity. Management Science **36/4**, 449–466 (1990)
75. Netessine, S., Dobson, G., Shumsky, Robert A.: Flexible Service Capacity: Optimal Investment and the Impact of Demand Correlation. Operations Research **50/2**, 375–388 (2002)
76. Fayezi, S., Zutshi, A., O'Loughlin, A.: International Journal of Management Reviews **19/4**, 379–407 (2017)

Chapter 2
Heuristics-based on the Hungarian Method for the Patient Admission Scheduling Problem

Rahma Borchani, Malek Masmoudi, Bassem Jarboui, and Patrick Siarry

Abstract This chapter tackles a healthcare problem, known as the patient admission scheduling (PAS) problem. PAS is a combinatorial optimization problem that involves assigning patients to beds in specialized rooms and departments depending on their medical requirements as well as their wishes. We present two heuristics based on the Hungarian algorithm, one to solve the PAS problem and the other to solve a relaxed version of the problem. The proposed methods are tested on the standard benchmark datasets and produce competitive results when compared with other state-of-the-art approaches.

2.1 Introduction

Healthcare presents one of the important sectors in the developed world. Overall, there is a growing need to improve the performance of healthcare services motivated by the rising costs and the growth rate of the population. Providing a patient admission schedule with good quality is a complex task that must take into account several factors. Among these we can cite the arrival and discharge dates of the patient, the bed to which the patient is assigned, the level of expertise of assigned room and department, the required medical treatment, the needed medical

R. Borchani (✉)
Faculty of Economics and Management of Sfax, Laboratory of Modeling and Optimization for Decisional, Industrial and Logistic Systems (MODILS), University of Sfax, Sfax, Tunisia

M. Masmoudi
Faculté des Sciences et Techniques, Université de Lyon, Université Jean Monnet Saint-Etienne, Lyon, France
e-mail: malek.masmoudi@univ-st-etienne.fr

B. Jarboui
Emirates College of Technology, Abu Dhabi, United Arab Emirates

P. Siarry
University of Paris-Est Creteil, Creteil, France
e-mail: siarry@u-pec.fr

© Springer Nature Switzerland AG 2021
M. Masmoudi et al. (eds.), *Operations Research and Simulation in Healthcare*, https://doi.org/10.1007/978-3-030-45223-0_2

equipment, the desired room features, the room gender policy, and the hospital's limited resources.

In this work, we are interested in the operational level of the PAS problem that was originally defined by Demeester et al. [1]. They formalized this problem through massive discussions with several decision makers in different hospitals, considering that all admission and discharge dates are known in advance and exclude intensive care units and day clinics. Furthermore, Demeester developed a website [2] that contains the PAS benchmark instances and their description, the penalty calculation, the weights of the cost function as well as a validator solution (Java program). This program presents a valuable tool to verify and validate every proposed solution. The same version of this PAS problem has attracted the interest of a lot of researchers. Bilgin et al. [4] employed a hyper-heuristic approach based on the strategy used by Burke et al. [17]. They considered tabu search as a heuristic selection method, and they employed four strategies as move acceptance criteria notably (1) improving and equal, (2) only improving, (3) simulated annealing, and (4) great deluge. Demeester et al. [5] introduced a mathematical model to solve this scheduling problem. This model presents the starting point for solving the PAS problem with integer programming. However, their integer program failed to find a feasible solution after 1 h of processing time. The calculation time is extended to one week without reaching the best solution. Due to the complexity of the problem, the authors proposed an approximate approach. They presented a quick decision support system based on tabu search algorithm hybridized with both token ring and variable neighbourhood descent approach. Ceschia and Schaerf [6] presented both simulated annealing and tabu search to address the PAS problem. They suggested a multi-neighbourhood local search procedure with different neighbourhood combinations. Besides, they proposed four lower bounds based on the relaxation of some constraints. Also, they tested their method on two instances with higher bed occupancy. Finally, they focused on the dynamic case, in which patients' admission and discharge become known during the planning period and not in advance. Bilgin et al. [7] focused on two timetabling problems in healthcare: the PAS problem and the nurse rostering problem. They proposed 36 hyper-heuristic variations based on the one described in Özcan et al. [18]. Numerical results show that the performance of the hyper-heuristic depends on the performance of the selection mechanism combined with the performance of the move acceptance criterion. Hammouri and Alrifai [8] considered the biogeography-based optimization approach to address the PAS problem. However, they stated that the proposed method needs further investigation in order to improve the quality results. Kifah and Abdullah [9] proposed an adaptive non-linear great deluge algorithm to solve the PAS problem. They tailored the classical great deluge algorithm by modifying the water level decay rate to a non-linear function. They increased the variations of the generated solutions using four variations of the neighbourhood structures in the search space. And they updated, in an adaptive manner, the water level decay rate. Turhan and Bilgen [10] solved the PAS problem via mixed integer programming based on two heuristics called fix-and-relax (F&R) and fix-and-optimize (F&O). The F&R heuristic provides feasible initial solutions in short calculation times, whereas the F&O heuristic improves this solution. In this work, authors focused on decomposition approaches. They decomposed patients based on their length of stays and preferences. Also, they decomposed the planning horizon into different optimization window sizes. Doush et al. [11] adapted the harmony search (HS) algorithm for solving the PAS problem. They distributed the patients randomly in the memory of harmony while respecting the feasibility of the obtained solution. They embedded five different neighbourhood strategies within the component of HS to explore more solutions in the search space. Bolaji et al. [12] proposed the late acceptance hill climbing (LAHC) algorithm as an iterative search method for solving the PAS problem. They used the room-oriented approach to generate a feasible initial solution. They proposed three neighbourhood structures to improve the quality of the initial solution, and they stored different accepted solutions in the table arrays to avoid being trapped in local minima. Range et al. [13] presented a slightly different mathematical model compared to Demeester et al. [5]. They proposed an optimization-based heuristic building on branch-and-bound, column generation, and dynamic constraint aggregation to solve the new variant of the PAS problem. Other researchers attempted to solve the PAS problem under uncertainty. Ceschia and Schaerf [14] introduced a new problem formulation of the dynamic patient admission scheduling problem under uncertainty (PASU). This problem considers registration dates, delayed admissions, emergency patients, uncertainty on the patients' length of stay, as well as overcrowding. Moreover, authors proposed a large set of new instances. They solved the problem using a simulated annealing approach. They suggested a local search procedure with different neighbourhood combinations. Ceschia and Schaerf [19] revisited the dynamic version of the PAS problem and considered constraints on the operating room utilization for patients needing surgery. Vancroonenburg et al. [31] studied PASU problem by introducing two ILP models. The first model considers only new patient arrivals, whereas the second model considers future planned arrivals. Both models are tested on a set of benchmark instances originally presented by Demeester et al. [5] and then extended to study the effect of uncertainty. They consider uncertainty on the patients' length of stay, as well as emergency patients. Zhu et al. [33] studied the compatibility of short-term and long-term objectives in the context of the dynamic PAS problem. They introduced two MIP formulations to solve short-term problems. Furthermore, they used Dantzig–Wolfe decomposition and column generation to calculate new lower bounds. In other studies, authors were particularly interested in the Patient Bed Allocation (PBA) problem. Schäfer et al. [30] introduced a Mixed Integer Programming (MIP) and a Greedy Look-Ahead heuristic (GLA heuristic) for solving the PBA problem in German hospital. They tested the proposed GLA heuristic with the literature data as well as real data from a hospital in Germany. Taramasco et al. [32] developed an autonomous bat algorithm for solving the PBA problem. To evaluate the proposed algorithm, experimental work was carried out on several instances taken from various hospitals and medical centres. The complexity of the underlying problem is proven as NP-complete [3], which implies that it cannot be solved in polynomial time. This encourages many researchers to apply approximate methods to solve this problem. In this chapter, we proposed two heuristics based on the Hungarian algorithm to solve the PAS problem and a relaxed version of the problem. The Hungarian algorithm was chosen for its good performances presented in the combinatorial optimization problem such as the incremental assignment problem [20], the fuzzy assignment problem [21], the assignment problem with changing costs[22], the 1–M assignment problem [23], the M–M assignment problem [24], the set cover problem [25], the no-wait flow shops [26], the computer vision technology [27], and the sequential ordering problem [28].

To evaluate the proposed heuristics, experimental work was carried out in twelve benchmark instances presented by Demeester et al. [5]. All of the empirical results and analysis illustrate that our approach obtains competitive results when compared with the existing approaches tackling the same problem in the literature.

The remainder of the chapter is organized as follows: Sect. 2.2 provides the definition of the PAS and its mathematical formulation. Section 2.3 describes our solution techniques. Benchmark instances, experimental results, and comparative analysis are discussed and presented in Sect. 2.4. Finally, the conclusion is given in Sect. 2.5.

2.2 PAS Problem Description and Formulation

The PAS problem can be described as follows: for a given planning horizon, a set of patients need to be assigned to a set of hospital beds for each night of their stay. Patients are identified through admission and discharge dates. These dates cannot be changed, and the length of stay must be contiguous. In addition, patients are characterized by their age, gender, specialism, room preference, and a set of necessary and preferred room properties. The hospital has several departments; each department is qualified to treat a set of specialisms at different levels of expertise. Some departments set a minimum or maximum age limit for admitted patients such as paediatrics that requires a maximum age limit of 16 years while gerontology requires a minimum age limit of 65 years. Each department has several rooms, and each room is qualified to treat a set of specialisms, each with its level of expertise. Besides, each room has different properties such as ventilation machine, telemetry, oxygen, and TV. So, each patient expresses these preferred properties in the assigned room. In addition to that, some specialisms might require the presence of certain specific properties in the room that the patient is going to occupy. These specific properties are presented as the patient's needed room properties. Furthermore, there are four different gender policies, presented by these letters F, M, D, N. Each room has a specific gender policy. Rooms of policy F indicate that only female patients are allowed. Rooms of policy M indicate that only male patients are allowed. Rooms of policy D indicates that both genders are allowed, but in the same room and every day the patients must have the same gender. So, the gender of the room is related to the gender of the first patient assigned to this room. Rooms of policy N indicates that mixed patients are allowed. Finally, each room is characterized by capacity, it can contain one, two, or four beds with the same characteristics. Each patient indicates his preference for the capacity of the assigned room he is going to occupy.

Figure 2.1 further illustrates the problem. Where, on the right side, we present some patients. Each patient is characterized by his age (PA), gender (PG), admission date (DA), and discharge date (DD). In addition, his specialism (PS), room capacity preference (RP), and a set of necessary and preferred room features (RFn, RFd). And, on the left side, we find the representation of the hospital, the hospital containing several departments (Dep1, Dep2, Dep ...). The first department sets a maximum age limit for admitted patients (Min_age, Max_age). It accepts patients under 65 years old. In addition, it is qualified to treat specialty 1 (s_1) with a high level of expertise (Major_S) and specialties 2 and 3 (s_2, s_3) with a low level of expertise (Minor_S). This department contains three rooms (r_1, r_2, and r_3). Each room is characterized by capacity (RC). For example, r_1 contains one bed (b_1), r_2 contains two beds (b_2, b_3), and r_3 contains four beds (b_4, b_5, b_6, b_7). All beds in the same room have the same characteristics. In addition, each room has a specific gender policy (RG), Major_S, Minor_S, and a set of room properties (RP).

The PAS problem consists in assigning a bed to each patient on each day of her/his stay period [6]. The assignment is subject to the following hard and soft constraints:

- Room capacity (RC): For each day, the number of patients admitted to a room must respect its capacity.
- Room gender (RG): For each day, patients affected in the same room must respect the gender policies of this room.
- Patient age (PA): Each patient should be assigned to the age-appropriate department during their stay.
- Room feature needed (RFn): Some patients' medical treatment requires a room with special properties during their stay.
- Room preference (RP): Each patient should be assigned to a room of his preferred capacity during his stay.
- Department specialism (DS): Each patient should be assigned to a department that offers the highest level of expertise for the specialism of the patient during his stay; a lower level of expertise is penalized.
- Room specialism (RS): Same as DS.
- Room features desired (RFd): Each patient should be assigned to a room that offers his desired features during his stay; every missing preferred property is penalized.
- Transfer (Tr): Each patient must not change his affected bed during his stay; every transfer is penalized.

The constraint RC presents a hard constraint, and then its satisfaction is obligatory in order to ensure a feasible solution. All other constraints are considered soft, and their satisfaction determines the quality of the solution. All soft constraints contribute, with their weights, to the objective function. So, their violations must be reduced to a minimum. The decision maker can assign these weights based on

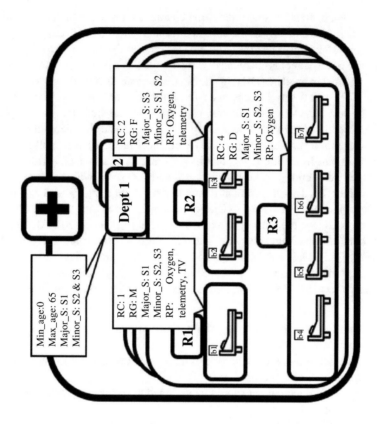

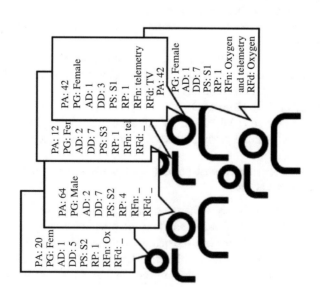

Fig. 2.1 Example of PAS problem

the hospital's internal rule, or specific hospital situations. Each weight assigned to a constraint determines its importance.

Ceschia and Schaerf [6] have merged the soft constraints PA, RFn, RP, DS, RS, RFd, and RG (if the room gender policy is only F or M) into a new constraint called patient-room cost (PRC) and also $C_{p,r}$ presents the total penalty of assigning a patient p to room r. We introduce here the mathematical model of the reformulated problem as it is presented in [6]. First, we present the symbols and the decision variables used to formulate the PAS mathematical model as follows: P presents the set of patients, B presents the set of beds, R presents the set of rooms, C_r presents the capacity of rooms $r \in R$, $P_f \in P$ presents the set of female patients, and $P_m \in P$ presents the set of male patients. D presents a set of days. D_p presents the set of days in which the patient $p \in P$ is present in the hospital, and $R_d \in R$ presents the set of rooms that have policy gender D.

The PAS decision variables are the following:

- $x_{p,\,r,\,d}$: 1 if patient p is assigned to room r in day d, 0 otherwise.
- t_p, r, d: 1 if patient p is transferred from room r in day d, 0 otherwise.
- f_r, d: 1 if there is at least one female patient in room r in day d, 0 otherwise.
- m_r, d: 1 if there is at least one male patient in room r in day d, 0 otherwise.
- b_r, d: 1 if there are both male and female patients in room r in day d, 0 otherwise.

The objective function, presented in Eq. (2.1), consists of minimizing the total penalties of the three soft constraints: PRC, TR, and RG (for room gender policy D):

$$F = F_{PRG} + F_{TR} + F_{RG}, \tag{2.1}$$

$$F_{PRG} = \sum_{p \in P, r \in R, d \in D} C_{p,r}\, x_{p,r,d}, \tag{2.2}$$

$$F_{TR} = \sum_{p \in P, r \in R, d \in D} w_{Tr}\, t_{p,r,d}, \tag{2.3}$$

$$F_{RG} = \sum_{r \in R_d, d \in D} w_{RG}\, b_{r,d}. \tag{2.4}$$

The hard and soft constraints are the following:

$$\sum_{r \in R} x_{p,\,r,\,d} = 1 \qquad \forall p \in P,\ d \in D_p \tag{2.5}$$

$$\sum_{p \in P} x_{p,\,r,\,d} \leq c_r \qquad \forall r \in R,\ d \in D \tag{2.6}$$

$$f_{r,d} \geq x_{p,\,r,\,d} \qquad \forall p \in P_f, r \in R,\ d \in D \tag{2.7}$$

$$m_{r,d} \geq x_{p,\,r,\,d} \qquad \forall p \in P_m r \in R,\ d \in D \tag{2.8}$$

$$b_{r,d} \geq m_{r,d} + f_{r,d} - 1 \qquad \forall r \in R, d \in D \qquad (2.9)$$

$$t_{p,r,d} \geq x_{p,r,d} - x_{p,r,d+1} \qquad \forall p \in P, r \in R, d \in D \qquad (2.10)$$

The constraint (2.5) ensures that every patient is assigned to a room during his stay. The constraint (2.6) ensures that the number of patients assigned to a room in each day cannot exceed its capacity. The constraint (2.7) reflects the existence of female in each room. Similar to the constraint (2.8) that reflects the existence of male existence in each room. The constraint (2.9) reflects the existence of both female and male in each room with gender policy D. If there are any, the total room gender penalty FRG is calculated. FRG is presented in constraint (2.4). The constraint (2.10) indicates whether each patient has been transferred from his room. In case the patient room is changed the total transfer penalty FTr is calculated. FTr is presented in constraint (2.3). Constraint (2.2) presents the total penalty of assigning patients to rooms during their stay.

2.3 Technical Solutions

Our technical solutions are based on the Hungarian method. This method is dedicated to solving the one-to-one (1–1) assignment problem, and it forces every patient to be assigned to exactly one bed. As a consequence, the hard constraint RC is satisfied. But it is difficult, if not impossible, to produce a feasible PAS solution that satisfies all the soft constraints due to the rugged search space and the highly critical nature of the problem [3]. Therefore, given that Tr constraint is the most crucial one [6], we decide to relax it and experiment the problem by canceling its penalty. In the following, we introduce the Hungarian method, we present an illustration on the PAS problem, and then we describe the technical solutions proposed for the relaxed and the global PAS problem.

2.3.1 Hungarian Method to PAS Problem

The Hungarian method (known as Kuhn–Munkres (K–M) algorithm) is a famous and traditional process used to deal with the assignment problems [24]. This combinatorial optimization algorithm can solve the linear assignment problem in $O(N^3)$ run time [15]. This method is originally developed by Harold Kuhn in 1955 [16] and refined by James Munkres in 1957 [15]. Basically, the Hungarian method operates on a principle of matrix reduction. Besides, it requires the existence of an N*N cost matrix as an input. The Hungarian method is an iterative procedure that transforms the cost matrix by subtracting and adding appropriate numbers until the best or least cost assignment is found. To explain the application of the Hungarian algorithm to the PAS problem, we present a small example of ten patients, one

Table 2.1 Patient details for the PAS example

P_Id	PA	PG	AD	DD	PS	LOS	RP	RFn	RFd
w_1	42	F	1	3	s_1	2	1	Oxygen and telemetry	None
w_2	20	F	1	2	s_2	1	1	Oxygen	None
w_3	49	F	1	3	s_3	2	1	None	Oxygen and telemetry
w_4	71	F	1	2	s_3	1	4	None	Telemetry
m_5	57	M	1	2	s_3	1	4	None	None
w_6	29	F	1	3	s_3	2	4	None	Oxygen and telemetry
m_7	62	M	1	3	s_2	2	4	None	Oxygen and telemetry
m_8	46	M	2	3	s_1	1	1	None	None
m_9	29	M	2	3	s_1	1	2	Oxygen and telemetry	None
w_{10}	16	F	2	3	s_2	1	2	None	Oxygen

Table 2.2 Department details for the PAS example

Department	Min_age	Max_age	Major_S	Minor_S
d_1	None	None	s_1	s_2 and s_3

Table 2.3 Room details for the PAS example

R_ID	RC	RG	Minor_S	Major_S	RP	LOB
r_1	1	D	s_1	s_2 and s_3	Oxygen and telemetry	b_1
r_2	2	D	s_3	s_1 and s_2	Oxygen and telemetry	b_2 and b_3
r_3	4	D	s_1	s_2 and s_3	Oxygen	$b_4, b_5, b_6,$ and b_7

department, three rooms, seven beds, and a planning horizon of two nights as shown in Tables 2.1, 2.2, and 2.3. Table 2.1 shows the details of each patient in our example. Column "P_Id" presents the patient identifier. To differentiate between women and men, we attribute the letter "w" to the women's identifier and the letter "m" to the men's identifier. Columns "PA" and "PG" show the age and the gender of each patient. Columns "AD" and "DD" present the admission and discharge date. "PS" presents the specialism of each patient. "LOS" presents the length of stay in the hospital. "RP" presents the capacity of the room that the patient prefers. "RFn" and "RFd" present the room features needed and desired.

In our example, the patient with the identifier w_1 is a 42-year-old woman. She should be assigned to a department that treats specialism s_1 for two nights. She prefers a single room and she needs oxygen and telemetry in the room.

Table 2.2 presents the details of the departments. In our instance, we have one department d1. It accepts patients of different ages. Specialism s_1 is the major specialism, and it is treated with a high level of expertise, differently to the two minor specialism s_2 and s_3, which are treated with a lower level of expertise than s_1.

Table 2.3 shows the details of each room in our example. Column "R_Id" presents the room identifier. Column "RC" presents the capacity of the room. Column "RG" shows the room gender policy. Columns "Minor_S" and "Major_S" present the level of expertise of each room for each specialism. Column "LOB" presents the list of beds in each room.

d_1	b_1	b_2	b_3	b_4	b_5	b_6	b_7
w_1	0	28	28	108	108	108	108
w_2	10	18	18	68	68	68	68
w_3	10	8	8	58	58	58	58
w_4	10	0	0	30	30	30	30
m_5	10	0	0	10	10	10	10
w_6	10	0	0	50	50	50	50
m_7	10	10	10	50	50	50	50

(a)

d_2	b_1	b_2	b_3	b_4	b_5	b_6	b_7
w_1	0	28	28	108	108	108	108
w_3	10	8	8	58	58	58	58
w_6	10	0	0	50	50	50	50
m_7	10	10	10	50	50	50	50
m_8	0	28	28	8	8	8	8
m_9	0	20	20	108	108	108	108
w_{10}	10	10	10	38	38	38	38

(b)

Fig. 2.2 The original cost matrices of the PAS example. (**a**) The original cost matrix of day 1. (**b**) The original cost matrix of day 2

In our example, room r_2 has a capacity of 2 beds (b_2 and b_3). Its gender policy is D. It can accept patients with specialism s_1, s_2, and s_3. Room r_2 has a high level of expertise for specialism s_3 and a lower level for specialism s_1 and s_2. This room contains oxygen and telemetry as available equipment.

For simplicity of the example, we choose to include only soft constraints PA, RFn, RP, DS, RS, RFd, and RG (if the room gender policy is only F or M). Their penalties are calculated in a single cost matrix called the patient-bed penalty matrix PB. This matrix is computed once for each day of the planning horizon, and the penalty calculation formulas are defined on the website [2]. Figure 2.2 below shows the original cost matrices of the first and second days of the example. In which, each of the seven patients (w_1, w_2, w_3, w_4, m_5, w_6, and m_7) present in the first day and each of the seven patients (w_1, w_3, w_6, m_7, m_8, m_9, and w_{10}) present in the second day need to be assigned in the seven beds (b_1 to b_7), one patient per bed.

The Hungarian method transforms the cost matrix iteratively to a sequence of equivalent matrices until finding the most favorable solution. The different steps of the application of the Hungarian method to the PAS problem are explained in the index with an illustrative example on the matrix cost of the first day. Figure 2.3 shows the outputs of the Hungarian method applied to our example. Figure 2.3a corresponds to the optimal assignment of day d1. Thus, patient w_1 should be assigned to bed b_1, patient w_2 bed b_3, patient w_3 bed b_2, patient w_4 bed b_5, patient m_5 bed b_4, patient w_6 bed b_6, and patient m_7 should be assigned to bed b_7. The total cost of this optimal assignment is $0 + 18 + 8 + 30 + 10 + 50 + 50 = 166$. Figure 2.3b corresponds to the optimal assignment of day d2. Thus, patient w_1 should be assigned to bed b_1, patient w_3 bed b_6, patient w_6 bed b_3, patient m_7 bed b_5, patient m_8 bed b_4, patient m_9 bed b_2, patient w_{10} should be assigned to bed b_7. The total cost of this optimal assignment is $0 + 58 + 0 + 50 + 8 + 20 + 38 = 174$ and the total cost of the example is $166 + 174 = 340$.

w_1	w_2	w_3	w_4	m_5	w_6	m_7	$f_{eval}*$
b_1	b_3	b_2	b_5	b_4	b_6	b_7	166

(a)

w_1	w_3	w_6	m_7	m_8	m_9	w_{10}	$f_{eval}*$
b_1	b_6	b_3	b_5	b_4	b_2	b_7	174

(b)

Fig. 2.3 The outputs of the Hungarian method applied to the PAS example. (a) Solution's presentation of the first day. (b) Solution's presentation of the second day

2.3.2 Heuristic1 for Relaxed PAS Problem

In this case, we set the Tr constraint's weight value at zero and keep the remaining constraints PA, RFn, RP, DS, RS, RFd, and RG at their predefined weight values. The Tr penalty associated with each patient depends on his assignment on the last day. This dependency makes finding a high-quality assignment more complex [6]. When we eliminate the penalty of the constraint Tr, we eliminate the correlation between assigned beds during the planned stay of each patient. In such a case, every single day could be scheduled independently and efficiently. Therefore, we propose heuristic1 based on the Hungarian method to solve this relaxed problem. The pseudo-code of this heuristic is shown in Algorithm 1. As in the previous Sect. 2.3.1, the penalties of the soft constraints PA, RFn, RP, DS, RS, RFd, and RG (if the room gender policy is only F or M) are calculated in the cost matrix PB. After applying the Hungarian method and based on the proposed solution, we calculate the penalty RG (if the room gender policy is D) as shown in Algorithm 2. Then, we evaluate this solution as shown in Algorithm 3. These steps are repeated for each day of the planning horizon until all rooms are filled with only men or only women, or until a maximum number of iterations is reached.

Algorithm 1 Heuristic1 to solve the relaxed PAS problem

Initialize the patient-bed penalty matrix PB;
Initialize the dominate room gender matrix drg;
Let f^*_{eval} the cost of the best solution and x the solution' vector;
Let F_{EVAL} the total cost and X the solution' matrix;
begin
    for each day d in the planning horizon PH **do**
        set f^*_{eval} = big number;
        while termination criteria is not met **do**
            x= Hungarian_method($PB[d]$);
            RGD_Penalty($PB[d],x,drg[d]$);
            Evaluate_solution+(x,f^*_{eval});
        $X[d]$= x^*; /*we save the best solution of each day*/
        F_{EVAL}+= f^*_{eval};

The pseudo-code to calculate the penalty RG (if the room gender policy is D) is shown in Algorithm 2: we compare the total number of men and women assigned in each room. If the number of women is greater than the number of men, we seek to eliminate all other men from this room. Therefore, we penalize the men of this room. Then, we verify the old dominate room gender; if drg = 'M', it means that in the previous step the women of this room have been penalized. So, we subtract the penalty w_{RG} from the assignment's cost of each woman and we define the drg as female, drg = 'F'. If the number of men is greater than the number of women, w_{RG} is added to the assignment's cost of each woman in the room. We update the assignment's cost of each man if the old drg is 'F'. Then, we define the new drg of this room as men, drg = 'M'. We repeat these steps for each room r with gender policy D.

Algorithm 2 RGD_Penalty(PB: penalty matrix, x: solution vector, drg: vector)

---

for each room r with gender policy D **do**
    Let $drg[r]$ the gender policy that dominates in the room r;
    if the number of women in r > the number of men in r **then**
        Add w_{RG} to the assignment' cost of each man in this room;
        if the old $drg[r]$ = 'M' **then**
            └ substract w_{RG} from the assignment' cost of each woman of this room;
        └ $drg[r]$ = 'F';
    else
        Add w_{RG} to the assignment' cost of each woman in this room r;
        if the old $drg[r]$ = 'F' **then**
            └ substract w_{RG} from the assignment' cost of each man of this room;
        └ $drg[r]$ = 'M';
Output: PB, drg;

The pseudo-code to evaluate the solution is shown in Algorithm 3: we calculate the total cost f_{eval} of the solution x. If this new solution offers superior quality compared to the best solution found so far, the new candidate is stored as the best solution.

Algorithm 3 Evaluate_solution(x: solution vector, f^*_{eval}: variable)

Calculate the cost function feval;
if $f_{eval} < f^*_{eval}$ **then**
    | $x^* = x$; $f^*_{eval} = f_{eval}$;
end
Output: x^*, f^*_{eval};

We apply heuristic1 to the previous example presented in Tables 2.1, 2.2, and 2.3. We start with the first day, and we consider the solutions shown in Fig. 2.3a as the output of the Hungarian method. We observe that beds b_4, b_5, b_6, and b_7 belonging to r_3 contain two women (w_4 and w_6) and two men (m_5 and m_7). Then, room 3 does not respect the constraint RG, and the total cost increases from 166 to 266. The steps from l_8 to l_{12}, described in heuristic1, are repeated until the best solution is found. This solution respects the constraint RG and reduces the total cost of the first day from 266 to 216. We repeat the same process to the second day in which r_2 and r_3 do not respect the constraint RG. The total cost of the second day reduces from 324

w_1	w_2	w_3	w_4	m_5	w_6	m_7	f_{eval}^*
b_1	b_7	b_5	b_4	b_2	b_6	b_3	216

(a)

w_1	w_3	w_6	m_7	m_8	m_9	w_{10}	f_{eval}^*
b_1	b_4	b_5	b_2	b_7	b_3	b_6	284

(b)

Fig. 2.4 Solutions' presentations of the example applied to the relaxed PAS problem. (**a**) Solution's presentation of the first day. (**b**) Solution's presentation of the second day

to 284, and the total cost of the solution reduces from 590 to 500. The best solutions x^* with their cost f_{eval}^* are shown in Fig. 2.4.

2.3.3 Heuristic2 for PAS Problem

In this case, we propose heuristic2 based on the Hungarian method to solve the PAS problem. This heuristic includes all soft constraints with their default weight values. Its pseudo-code is shown in Algorithm 4. The principle of the heuristic2 is to add the costs of the TR and RG constraints to the penalty matrix of each day before solving it. Based on the solution proposed by the Hungarian method, we update the dominate gender in each room. Consequently, we update the penalties due to violation of the constraint RG, and finally we evaluate the new solution. We repeat these instructions starting from the first to the last day in the planning horizon, and then we repeat the same steps starting from the last to the first day, which gives us a wide and flexible search space. These steps are detailed in Algorithm 5.

To calculate the RG penalty (if the room type rule is D), we use the same method as in Algorithm 2, with some modifications in lines 1_4 and 1_6. Before adding the penalty w_{RG} to the cost of assigning each patient to a room in 1_4, you must multiply this penalty with the number of the remaining days in which the patient will occupy the bed. The same value should be subtracted from 1_6. For example, in the case of two men and only two free beds. The first man stays for two nights, and the second one stays for six nights. The first bed is located in a room that contains 3 women, and the second bed is located in a room that contains 3 men. So, we have to assign one man to a women's room and the other to a men's room, but who will be assigned to the women's room and who will be assigned to the men's room ?

As a solution, we multiply the penalty with the length of stay of each man and we choose to affect the man with the lowest penalty in the women's room. In our example, we affect man 1 to the women's room because his penalty $(2 * w_{RG})$ will be weaker than the penalty of affecting the second man in this room, which is equal to $(6 * w_{RG})$. The multiplication of the weight of the constraint RG by the number of the remaining days leads to privileging the satisfaction of this constraint comparing to the constraint TR. So, to balance these two constraints, we multiply the weight of the constraint Tr by a coefficient $coef_{Tr}$. As shown in the Algorithm 6, we add this new penalty ($coef_{Tr} * w_{Tr}$) to the cost of assigning each patient to each bed different from its affected bed in the last day.

Algorithm 4 Heuristic2 to solve the PAS problem

Initialize the patient-bed penalty matrix PB;
Initialize the dominate room gender matrix drg;
Initialize the vector of the average length of patients' stay in each day Alos;
Let f_{eval}^* the cost of the best solution in each day;
begin
    $Ctr = 0$;
    while termination criteria is not met **do**
        for $d = 0 .. PH$ **do**
            Hungarian_RGD_TR($PB[d], x[d], drg[d], x[d-1], x[d+1]$,
                    $drg[d-1], 0, ctr, f_{eval}^*$);
        ctr++;
        for $d = PH .. 0$ **do**
            Hungarian_RGD_TR($PB[d], x[d], drg[d], x[d+1], x[d-1]$,
                    $drg[d+1], PH, ctr, f_{eval}^*$);

Algorithm 5 Hungarian_RGD_TR(PB: penalty matrix of day d, x: solution vector of day d, drg: dominate room gender vector of day d, $x1$: vector, $x2$: vector, $drg1$: vector, dd: variable, ctr: variable

---

/* if the boucle is from $d = 0..PH$*/
Let $x1$ the solution vector of day $d-1$ and $x2$ the solution vector of day $d+1$;
Let $drg1$ the vector of the dominate room gender of day $d-1$; Let dd be 0;
/* if the boucle is from $d = PH..0$*/
Let $x1$ the solution vector of day $d+1$ and $x2$ the solution vector of day $d-1$;
Let $drg1$ the vector of the dominate room gender of day $d+1$;
Let dd be PH;
Set $coef_{TR}$ a random real number between 0.1 and $Alos$;
if $d \neq dd$ **then**
    RGD_Penalty($PB, x, drg1$);
    Tr_Penalty($PB, x1, coef_{TR}$);
    if $ctr > 0$ **then**
        Tr_Penalty($PB, x2, coef_{TR}$);
    end
end
$x=$ Hungarian_method(PB);
RGD_Penalty(PB, x, drg);
Evaluate_solution(x, f_{eval}^*);
Output: $x*, f_{eval}^*$

w_1	w_2	w_3	w_4	m_5	w_6	m_7	f_{eval}^*
b_1	b_5	b_7	b_4	b_2	b_6	b_3	216

(a)

w_1	w_3	w_6	m_7	m_8	m_9	w_{10}	f_{eval}^*
b_1	b_7	b_6	b_3	b_5	b_2	b_4	234

(b)

Fig. 2.5 Solutions' presentations of the example applied to the PAS problem. (**a**) Solution's presentation of the first day. (**b**) Solution's presentation of the second day

---

Algorithm 6 Tr_Penalty (PB: penalty matrix, x: solution vector, $coef_{TR}$: real variable)

---

for each bed b in the solution x **do**
  let p the patient affected to bed b;
  for each bed bd in the list of beds **do**
    if $bd \neq b$ **then**
      $PB[p, bd] \mathrel{+}= coef_{Tr} * w_{Tr}$;
    end if
  end
end
Output PB

We apply heuristic2 to our example. we notice that the solution proposed by the Hungarian method, shown in Fig. 2.3, does not respect the constraint RG, as has already been described in Sect. 2.3.2. Furthermore, this solution does not respect the constraint TR. In which patient w_3 transfers from bed b_2 to bed b_6, w_6 transfers from bed b_6 to bed b_3, and m_7 transfers from bed b_7 to bed b_5. In calculating the penalties of the RG (250) and TR (330) constraints, the total cost of this solution will be 920. After applying heuristic2, total cost of the solution reduces from 920 to 450. The best solutions x* with their cost f_{eval}^* are shown in Fig. 2.5.

2.4 Experimental Results and Discussions

In this section, we first introduce the benchmark instances, then we show our experimental results, and finally we propose a comparative analysis with previous work.

2.4.1 Benchmark Instances and Setting

This work focuses on experimenting 12 out of the 13 benchmark instances presented by Demeester et al. [1, 5] and available on his website [2] using the same constraint's weights they have defined. The generation of these benchmarks is based on a realistic hospital situation and on discussions with experienced decision makers

Table 2.4 PAS instances

	Instances	P	D	S	B	R	RP	PH	ALOS	ACPR
Small instances	1	652	4	4	286	98	2	14	3.66	32.16
	2	755	6	6	465	151	2	14	5.17	36.74
	3	708	5	5	395	131	2	14	4.46	35.96
	4	746	6	6	471	155	2	14	4.79	38.39
	5	587	4	4	325	102	2	14	3.82	31.23
	6	685	4	4	313	104	2	14	4.12	29.53
	Average	689	5	5	376	124	2	14	4.34	34.00
Large instances	7	519	6	6	472	162	4	14	4.27	102.87
	8	895	6	6	441	148	4	21	4.54	103.72
	9	1400	4	4	310	105	4	28	4.90	113.25
	10	1575	4	4	308	104	4	56	5.23	80.57
	11	2514	4	4	318	107	4	91	5.28	91.29
	12	2750	4	4	310	105	4	84	5.19	94.00
	Average	1609	5	5	360	122	4	49	4.90	97.62

Table 2.5 Constraints' weights

Constraint	Weight's symbol	Corresponding weight
RG	w_{RG}	5.0
PA	w_{PA}	10.0
RFn	w_{RFn}	5.0
RP	w_{RP}	0.8
DS	w_{DS}	1.0
RS	w_{RS}	1.0
RFd	w_{RFd}	2.0
Tr	w_{Tr}	11.0

in several hospitals [5]. The first 12 instances contain six small and six large datasets and involve patients with only one specialism, while instance 13 involves patients with a multi-specialism and has a different structure than the foregoing instances. That is why we exclude instance 13. Table 2.4 summarizes the main characteristics of all considered instances. Table 2.5 presents soft constraints and their corresponding weights.

For the small instances, the average number of patients (P) is 689, while the average number of patients is 1609 for large instances. The number of departments (D) as the number of specialism (S) varies between 4 and 6. Each department deals with three different specialisms, once as a major and twice as a minor. As it is explained in [5], the number of available rooms (R) in each department varies between 20 and 30. Each of these rooms has a number of beds (B), and they are distributed as follows:

- At most 5 rooms that contain a single bed,
- Between 5 and 10 that contain twin beds,
- And for the remaining rooms, they contain quadruple beds.

For the small instances, each room can have 0, 1, or 2 room properties (PR). Whereas, for the large, we can even find four different properties. The planning horizon (PH) is two weeks for the first six instances while, for the last one, it ranges from 14 to 91 days, which explains the increase of the average length of stay for all patients (ALOS). It can be observed from these different characteristics that the large instances have more heterogeneous rooms; the average number of patients, the average length of the planning horizon, and the average number of room properties for the large instances are more than twice compared with the small one, while the numbers of the other parameters (D, S, B, R) remain the same, which explains the big increase in the average cost of patient room assignment (ACPR) column in the large instances.

The proposed method is implemented in C++ language using the Microsoft Visual Studio 2017. The computational tests have been performed on a Windows 10 PC with an Intel Core i5 CPU (1.8 GHz) and 4 GB RAM. All the instances results have been validated using the Validator program available at the website [2].

2.4.2 Experimental Results

2.4.2.1 Experimental Results on Relaxed PAS Problem

The first set of experiments is performed to show the results of the relaxed problem by changing the weight of the transfer constraint w_{Tr} to zero and keeping all the other weights unchanged.

Ceschia and Schaerf [6] are the only researchers who have worked on the same relaxed problem. They have proposed a lower bound (LB) based on the relaxation of Tr constraint. In addition, they tested their SA approach assuming that $w_{Tr} = 0$.

The results of our approach are presented in Table 2.6 in comparison with the results of the two approaches proposed in [6]. The columns "Heuristic1" and "SA" present the best cost obtained for 10 runs.

The first outcome is that our results outperform eleven of twelve results giving by SA (see Fig. 2.6), with a positive percentage deviation presented in the column "$\%Gap_{SA}$", principally shown in the large instances.

The second outcome is that the percentage deviation between our results and the LB presented in the column "$\%Gap_{LB}$" is not significantly worse.

These outcomes ensure that our solution offers good quality results. Note that the percentage deviation is computed as follows:

$$Gap = 100((Another\ approach - Our\ heuristic)./Our\ heuristic) \quad (2.11)$$

Table 2.6 Comparison of our results with previous work for the PAS relaxed problem

Instances	Heuristic1	LB [6]	$\%Gap_{LB}$	SA [6]	$\%Gap_{SA}$
1	644.8	637.6	−1.12	647.7	0.45
2	1123.4	1104	−1.73	1116.2	−0.64
3	737.6	722.8	−2.01	744	0.87
4	1081.6	1074	−0.70	1117.5	3.32
5	621.6	618.4	−0.51	621.6	0.00
6	775.2	769.6	−0.72	780.4	0.67
7	1027.4	682.2	−33.60	1117.2	8.74
8	3705.6	2637	−28.84	3888	4.92
9	17122.4	10090	−41.07	20233.6	18.17
10	6954.2	6656	−4.29	7655.1	10.08
11	9540	7831	−17.91	10977.1	15.06
12	19706.8	12575	−36.19	22419.3	13.76

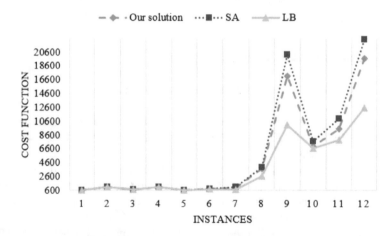

Fig. 2.6 Comparison of the results of the PAS relaxed problem

2.4.2.2 Experimental Results on PAS Problem

Our second set of experiments is performed with the default weighting values, which allow us to compare with the previous techniques that worked on the same formulation of this scheduling problem in the scientific literature. For small instances, we compare our results with thirteen previous techniques. Although, for large instances, there are only two techniques to be compared. These comparative techniques are summarized and abbreviated in Table 2.7.

The results of the small instances are given in Table 2.8. It accounts for each approach and for each instance "I": the best solution, "*Best*"; the average solution of 10 runs, "*Avg.*"; the standard deviation, "*St.dev.*", and the rank based on the average and the standard deviation presented as (*Avg./ St.dev.*). The rank (4/1) in instance 3 is relevant to the order of all the fourteen approaches, and it means that

Table 2.7 Key to the comparative techniques

Key	Approach name	Reference
HH-GD	Hyper-heuristic great deluge	[4]
HH-SA	Hyper-heuristic simulated annealing	[4]
HH-IE	Hyper-heuristic improving equal	[4]
TS	Tabu search	[4]
SA	Simulated annealing	[6]
DHS-SA	Dynamic heuristic set strategy with a simple random simulated annealing	[7]
DHS-OI	Dynamic heuristic set strategy with simple random-only improvement	[7]
DHS-GD	Dynamic heuristic set strategy with a simple random great deluge	[7]
BBO	Biography-based optimization	[8]
GD	Standard greet deluge	[9]
ANLGD	Adaptive non-linear great deluge	[9]
ANLGDH	Adaptive non-linear great deluge with hard constraint violation during the search	[9]
$MIP_{F\&R;F\&O}$	Mixed integer programming based on heuristics F&R and F&O	[10]
HS	Harmony search	[11]

the proposed method is ranked fourth based on the average results and it is ranked first based on the standard deviation.

Experimental results presented in Table 2.8 show that the proposed method achieves competitive, robust, and stable results and works comparatively well across all the small instances in comparison with the other thirteen approaches in the literature; it is ranked fifth in terms of average. However, it is ranked nearly the third in terms of standard deviation. The results of the big instances are given in Table 2.9. The numbers in this table report for each approach and for each instance: the best solution, "*Best*"; the average solution of 10 runs, "*Avg.*"; the gap, "*Gap_{best}*" is the percentage deviation between the best solutions of our results and the best-known solutions presented by Ceshia et al. [6]; the gap "*Gap_{avg}*" is the percentage deviation between the average of our results and the average results reported by the other two approaches. This percentage deviation is computed as mentioned in Eq. (2.11). We did not indicate the "*Best*" and "*Gap_{best}*" columns for the approach presented by Turhan and Bilgen [10] because no results were reported in their search. We have shown in the preceding subsection that large instances are difficult and heterogeneous, which explains why most of the researchers have experimented only the small instances whereas our approach can always produce solutions, even for the large one with a gap ranging from -2 to -19% compared to the best-known solutions.

We also statistically processed the performance of our proposed heuristic for the PAS problem. Figure 2.7 reveals the distributions of the experimental results of all

Table 2.8 Comparison of the results for the small instances

I		Heuristic2	HH-GD [4]	HH-SA [4]	HH-IE [4]	TS [4]	SA [6]	DHS-SA [7]	DHS-OI [7]	DHS-GD [7]	BBO [8]	GD [9]	ANLGD [9]	ANLGDH [9]	HS [11]
1	Best	716.4	-	-	-	-	659.2	680.0	884.4	674.8	1233.4	710.0	704.4	664.6	742.0
	Avg.	727.0	876.9	830.3	893.6	1051.0	665.6	709.2	959.3	685.6	1312.8	753.2	744.9	678.6	774.6
	St.dev.	6.9	20.5	18.8	46.1	48.1	3.2	26.6	51.2	7.4	59.5	68.6	33.5	6.0	23.6
	Rank	5/3	10/6	9/5	11/10	13/11	1/1	4/8	12/12	3/4	14/13	7/14	6/9	2/2	8/7
2	Best	1231.8	-	-	-	-	1143.6	1191.6	1548.8	1185.2	2027.0	1258.6	1250.8	1162.4	1273.6
	Avg.	1240.6	1475.8	1382.2	1525.1	2211.8	1150.9	1213.3	1622.8	1202.6	2088.9	1280.6	1275.3	1175.4	1328.2
	St.dev.	5.4	18.0	14.2	51.9	87.2	3.5	17.5	39.9	15.8	60.4	16.9	12.4	6.2	27.7
	Rank	5/2	10/9	9/5	11/12	14/14	1/1	4/8	12/11	3/6	13/13	7/7	6/4	2/3	8/10
3	Best	841.4	-	-	-	-	776.6	823.6	1126.0	803.8	1385.2	859.2	836.4	792.4	908.4
	Avg.	847.2	986.1	923.5	991.1	1184.2	786.6	847.2	1185.1	822.6	1433.7	900.2	891.9	804.7	939.4
	St.dev.	4.6	32.9	15.3	30.7	60.3	5.3	15.1	35.0	22.5	41.7	48.1	33.3	6.6	16.7
	Rank	4/1	10/9	8/5	11/8	12/14	1/2	5/4	13/11	3/7	14/12	7/13	6/10	2/3	9/6
4	Best	1282.6	-	-	-	-	1176.0	1274.6	1633.8	1228.2	2211.0	1339.8	1306.4	1187.4	1334.4
	Avg.	1330.6	1631.2	1624.0	1742.4	3663.4	1190.5	1310.4	1749.8	1251.0	2301.5	1367.5	1340.6	1219.9	1392.3
	St.dev.	22.2	34.1	38.1	34.4	163.4	4.3	22.0	67.5	12.2	69.9	23.6	31.5	17.6	41.6
	Rank	5/5	10/8	9/10	11/9	14/14	1/1	4/4	12/12	3/2	13/13	7/6	6/7	2/3	8/11
5	Best	668,8	-	-	-	-	625.6	640.8	749.4	639.2	800.8	650.4	656.0	634.4	668.0
	Avg.	671,7	688.6	661.6	684.4	714.1	631.8	648.0	771.4	642.8	828.9	663.4	677.4	635.8	680.5
	St.dev.	2,7	6.3	4.4	9.1	6.2	1.7	5.0	16.8	4.7	28.3	5.2	14.4	1.2	8.4
	Rank	7/3	11/9	5/4	10/11	12/8	1/2	4/6	13/13	3/5	14/14	6/7	8/12	2/1	9/10
6	Best	858	-	-	-	-	801.2	835.8	1073.6	825.8	1283.2	910.2	878.8	806.8	916.6
	Avg.	870,4	1034.9	955.0	1001.8	1188.1	811.1	855.5	1137.3	836.0	1317.1	931.1	900.6	815.4	935.2
	St.dev.	7,3	30.9	20.1	32.9	58.1	3.0	14.1	43.8	8.8	37.9	11.3	15.2	5.3	17.1
	Rank	5/3	11/10	9/9	10/11	13/14	1/1	4/6	12/13	3/4	14/12	7/5	8/7	2/2	8/8
	Avg.rank	5.1/2.8	10.3/8.5	8.5/6.3	10.6/10.1	13/12.5	1/1.3	4.1/6	12.3/12	3/4.6	13.6/12.8	6.8/8.6	6.3/8.1	2/2.3	8.3/8.6

Dashes represent that no results are reported by the studies

Table 2.9 Comparison of the results for the large instances

Instances	Heuristic2		SA [6]				$MIP_{F\&R;F\&O}[10]$	
	Best	Avg.	Best	Avg.	Gap_{best}	Gap_{avg}	Avg.	Gap_{avg}
7	1482.4	1502.1	1199	1216.3	−19.1	−19.0	1426.6	−5
8	4925.2	4957.2	4158.6	4191.7	−15.5	−15.4	4673.4	−5.73
9	22507.6	22515.2	21942	22052.8	−2.5	−2	−	−
10	9702.4	9744	8146.6	8261.2	−16	−15.2	−	−
11	14638	14643.2	12016.8	12105.7	−17.9	−17.3	−	−
12	26546.6	26600.6	23758.4	23968.8	−10.5	−9.8	−	−

instances using the box and whisker plot, from which we can see the differences in the skewness pattern for each data of the twelve instances.

We can observe based on Fig. 2.7 that the results in instances 1 and 3 are uniformly distributed over the median; then the distributions for these two instances are symmetric. In instances 2, 5, 6, 7, 8, 9, and 10, most of the results are condensed in the scale lower bound; this indicates that the distribution is right-skewed. In instances 4 and 12, the distributions of the results are skewed to the upper bound.

The box and whisker plots presented in Fig. 2.7 show that the spreading of the experimental solutions of the instances has different patterns. Therefore, we conclude that the behavior of our heuristics differs from one instance to another due to the difference in the instances' structure.

2.5 Conclusions

In this study, we developed two heuristics based on the Hungarian method to solve the PAS problem and a relaxed version of the problem. To test the performance of our heuristics, experiments were carried out in twelve instances.

For the relaxed problem, Heuristic1 outperformed eleven of twelve results giving by $SA_{wTr=0}$ with an average gap of 6.28%.

For the PAS problem, Heuristic2 provided competitive and comparative quality solutions principally shown in the small instances; it is ranked fifth when compared to other thirteen solution techniques in the scientific literature. For large instances, experimental results showed that the results correspond to an average gap of −14%. We observed that most research focuses on small instances, while our approach can always produce good solutions even for large instances.

For future research, we need to provide a sensitivity study to test all the parameters of our algorithms. Fuzzy study is also required to show the effectiveness of the Hungarian algorithm. The dynamic versions of the patient admission scheduling problem introduced by Ceschia and Schaerf [14, 19] are another interesting area of research aimed at integrating concrete situations such as the uncertainty of length of

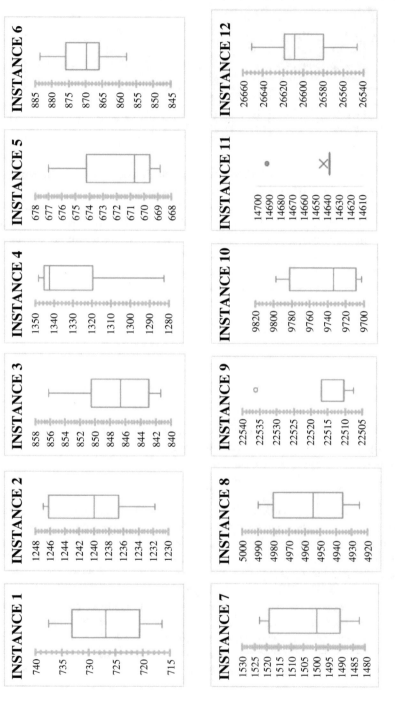

Fig. 2.7 The results' distributions for all the instances using the box and whisker plot

stay, the possibility to delay the admission of the patient, the risk of overcrowding, and the use of the operating room for patients requiring surgery.

Appendix

The Hungarian algorithm: An example

We present the three steps of the Hungarian method described in Shweta Singh et al. [29]. Step 0 is executed once, while steps 1 and 2 are repeated until an optimal assignment is found. The input of the algorithm is an n-by-n square cost matrix with only non-negative elements. We consider the example presented in Tables 2.1, 2.2, and 2.3. In which seven patients (w_1, w_2, w_3, w_4, m_5, w_6, and m_7) are present on the first day and should be assigned to seven beds (b_1 to b_7).

This is the original cost matrix:

d_1	b_1	b_2	b_3	b_4	b_5	b_6	b_7
w_1	0	28	28	108	108	108	108
w_2	10	18	18	68	68	68	68
w_3	10	8	8	58	58	58	58
w_4	10	0	0	30	30	30	30
m_5	10	0	0	10	10	10	10
w_6	10	0	0	50	50	50	50
m_7	10	10	10	50	50	50	50

Below we will show the different steps of the Hungarian algorithm and explain them using this example.

- **Step 0: Reduction on the matrix:**

  1. For each row on the cost matrix, find the lowest element and subtract it from each element in its row. As a result, we obtain at least one zero in this line. We find several zeros when there is more than one element equal to the smallest element in this line. The smallest element in the second row is, for example, 10. Therefore, we subtract 10 from each element in the second row. The resulting matrix is

0	28	28	108	108	108	108	
0	8	8	58	58	58	58	(-10)
2	0	0	50	50	50	50	(-8)
10	0	0	30	30	30	30	
10	0	0	10	10	10	10	
10	0	0	50	50	50	50	
0	0	0	40	40	40	40	(-10)

2. We do the same with the columns. For each column on the cost matrix obtained in part 1, find the lowest element and subtract it from each element in its column, giving the following matrix:

0	28	28	98	98	98	98
0	8	8	48	48	48	48
2	0	0	40	40	40	40
10	0	0	20	20	20	20
10	0	0	0	0	0	0
10	0	0	40	40	40	40
0	0	0	30	30	30	30
			(-10)	(-10)	(-10)	(-10)

- **Step 1: Cover all zeros with a minimum number of lines**

  1. Examine the rows successively until a row with the fewest unmarked zero is found. Make an assignment to one of these zeros by enclosing it in a box. Cross all other zeros appearing in the corresponding row or column; crossed zeros will not be considered for future assignment. Repeat this operation successively until one of the following situations arises:

     – All the zeros in rows/columns are either marked or crossed, and there is exactly one assignment in each row of the matrix, as well as in each column. In such a case, the algorithm stops, and an optimal solution for the given problem is obtained.
     – There may be some row (or column) without assignment, i.e., the total number of marked zeros is less than the order of the matrix. In such a case, we continue with these steps:

  2. Mark all rows without assignment.
  3. Mark all columns with a crossed zero in marked rows.
  4. Mark all rows that have assignments in marked columns.
  5. Repeat steps 2, 3, and 4 alternately until no more rows or columns can be marked.
  6. Draw straight lines through each unmarked row and each marked column.

If the number of drawn lines is equal to the order of the matrix, the algorithm stops, and the current solution presents the optimal solution. Otherwise, proceed to the next Step 2.

0	28	28	98	98	98	98	x
ө	8	8	48	48	48	48	x
2	ө	0	40	40	40	40	x
10	ө	ө	20	20	20	20	x
10	ө	ө	0	ө	ө	ө	
10	ө	ө	40	40	40	40	x
ө	ө	ө	30	30	30	30	x
x	x	x					

Because the number of drawn lines (4) is lower than the size of the matrix ($n = 7$), we continue with Step 2.

- **Step 2: Create additional zeros**

  1. Find the smallest uncovered number by a straight line.
  2. Subtract this element from all uncovered numbers by a straight line.
  3. Add this same smallest element to every number covered by two straight lines.

In our example, the number of lines is smaller than 7. The smallest uncovered number is 20. We subtract this number from all uncovered elements and add it to all elements that are covered twice:

First, we find that the smallest uncovered element is 20. We subtract this number from all uncovered numbers and add it to all numbers that are covered twice. This results in the following matrix:

0	28	28	78	78	78	78
0	8	8	28	28	28	28
2	0	0	20	20	20	20
10	0	0	0	0	0	0
30	20	20	0	0	0	0
10	0	0	20	20	20	20
0	0	0	10	10	10	10

Repeat steps 1 and 2 until there is one zero in each row and in each column of the matrix. The positions of the assigned zeros describe an optimal solution in the initial cost matrix.

- Step 1: Cover all zeros with a minimum number of line Again, we determine the minimum number of lines required to cover all zeros in the matrix. Now there are 5 lines required:

0	28	28	78	78	78	78	x
ө	8	8	28	28	28	28	x
2	ө	0	20	20	20	20	x
10	ө	ө	ө	ө	ө	0	
30	20	20	ө	ө	0	ө	
10	ө	ө	20	20	20	20	x
ө	ө	ө	10	10	10	10	x
x	x	x					

- Step 2: Create additional zeros The number of lines is smaller than 7. The smallest uncovered number is 10. We subtract this number from all uncovered elements and add it to all elements that are covered twice:

0	28	28	68	68	68	68
0	8	8	18	18	18	18
2	0	0	10	10	10	10
20	10	10	0	0	0	0
40	30	30	0	0	0	0
10	0	0	10	10	10	10
0	0	0	0	0	0	0

- Step 1: Cover all zeros with a minimum number of lines There are 6 lines required to cover all zeros:

0	28	28	68	68	68	68	x
ө	8	8	18	18	18	18	x
2	ө	0	10	10	10	10	
20	10	10	ө	ө	ө	0	
40	30	30	ө	ө	0	ө	
10	0	ө	10	10	10	10	
ө	ө	ө	ө	0	ө	ө	
x							

- Step 2: Create additional zeros The number of lines is smaller than 7. The smallest uncovered number is 8. We subtract this number from all uncovered elements and add it to all elements that are covered twice:

0	20	20	60	60	60	60
0	0	0	10	10	10	10
10	0	0	10	10	10	10
28	10	10	0	0	0	0
48	30	30	0	0	0	0
18	0	0	10	10	10	10
8	0	0	0	0	0	0

- Step 1: Cover all zeros with a minimum number of lines There are 6 lines required to cover all zeros:

0	20	20	60	60	60	60	x
⊖	⊖	0	10	10	10	10	x
10	⊖	⊖	10	10	10	10	x
28	10	10	⊖	⊖	⊖	0	
48	30	30	⊖	⊖	0	⊖	
18	⊖	⊖	10	10	10	10	x
8	⊖	⊖	⊖	0	⊖	⊖	
x	x	x					

- Step 2: Create additional zeros The number of lines is smaller than 7. The smallest uncovered number is 10. We subtract this number from all uncovered elements and add it to all elements that are covered twice:

0	20	20	50	50	50	50
0	0	0	0	0	0	0
10	0	0	0	0	0	0
38	20	20	0	0	0	0
58	40	40	0	0	0	0
18	0	0	0	0	0	0
18	10	10	0	0	0	0

- Step 1: Cover all zeros with a minimum number of lines There is exactly one assignment in each row, as well as in each column of the matrix. An optimal assignment exists among the zeros included in the boxes. Therefore, the algorithm stops.

  This corresponds to the following optimal assignment in the original cost matrix: This solution is summarized in the following vector in which the last box presents the total cost f_{eval}

0	20	20	50	50	50	50
0	0	0	0	0	0	0
10	0	0	0	0	0	0
38	20	20	0	0	0	0
58	40	40	0	0	0	0
18	0	0	0	0	0	0
18	10	10	0	0	0	0

0	28	28	108	108	108	108
10	18	18	68	68	68	68
10	8	8	58	58	58	58
10	0	0	30	30	30	30
10	0	0	10	10	10	10
10	0	0	50	50	50	50
10	10	10	50	50	50	50

w_1	w_2	w_3	w_4	m_5	w_6	m_7	f_{eval}
b_1	b_3	b_2	b_5	b_4	b_6	b_7	166

Patient 1 should be assigned to bed 1, patient 2 bed 3, patient 3 bed 2, patient 4 bed 5, patient 5 bed 4, patient 6 bed 6, and patient 7 should be assigned to bed 7. The total cost of this optimal assignment is $0 + 18 + 8 + 30 + 10 + 50 + 50 = 166$.

References

1. Demeester P, De Causmaecker P, Vanden Berghe G (2008) Applying a local search algorithm to automatically assign patients to beds. In: Proceedings of the 22nd conference on quantitative methods for decision making. Orbel22:35–36
2. Demeester P (2009) Patient admission scheduling website. https://people.cs.kuleuven.be/~wim.vancroonenburg/pas/. viewed 26November 2018
3. Vancroonenburg W, Goossens D, Spieksma F (2011) On the complexity of the patient assignment problem. Technical report, KAHO Sint-Lieven, Gebroeders De Smetstraat 1, Gent, Belgium.
4. Bilgin B, Demeester P, Vanden Berghe G (2008) A hyper-heuristic approach to the patient admission scheduling problem. Technical Report, KaHo Sint-Lieven, Gent.
5. Demeester P, Souffriau W, De Causmaecker P, Vanden Berghe G (2010) A hybrid tabu search algorithm for automatically assigning patients to beds. Artificial Intelligence in Medicine 48:61–70
6. Ceschia s, Schaerf A (2011) Local search and lower bounds for the patient admission scheduling problem. Comput Oper Res 38:1452–1463

7. Bilgin B, Demeester P, Misir M, Vancroonenburg W, Berghe G V (2012) One hyper-heuristic approach to two timetabling problems in health care J Heuristics 18:401–434
8. Hammouri A I, Alrifai B (20104) Investigating biogeography-based optimization for patient admission scheduling problems. J Theor Appl Inf Technol 70:413–421
9. Kifah S, Abdullah S (2017) An adaptive non-linear great deluge algorithm for the patient-admission problem. Inf Sci 295:573–585
10. Turhan A M, Bilgen B (2017) Mixed integer programming based heuristics for the patient admission scheduling problem. Comput Oper Res 80:38–49
11. Doush I A, Al-Betar M A, Awadallah M A,Hammouri A I, Al-Khatib R M, El Mustafa S, ALkhraisat H (2018) Harmony Search Algorithm for Patient Admission Scheduling. J Intell Syst
12. Bolaji A L, Femi B A, Shola P B (2018) Late acceptance hill climbing algorithm for Solving Patient Admission Scheduling Problem. Knowl-Based Syst 145:197–206
13. Range T M, Lusby R M, Larsen J (2014) A column generation approach for solving the patient admission scheduling problem. Eur J Oper Res 235:252–264
14. Ceschia S, Schaerf A (2012) Modeling and solving the dynamic patient admission scheduling problem under uncertainty. Artif Intell Med 56:199–205
15. Munkres J (1957) Algorithms for the assignment and transportation problems. J Soc Ind Appl Math 5:32–38
16. Kuhn H W (1955) The Hungarian method for the assignment problem. Nav Res Logist Q 2:83–97
17. Burke E, Kendall G, Newall J, Hart E, Ross P, Schulenburg S (2003) Hyper-heuristics: an emerging direction in modern search technology. Int Series Oper Res Manag Sci 457–474
18. Özcan E, Bilgin B, Korkmaz E E (2008) A comprehensive analysis of hyper-heuristics. Intell Data Anal 12:3–23
19. Ceschia S, Schaerf A (2014) Dynamic patient admission scheduling with operating room constraints, flexible horizons, and patient delays. J Sched 14:377–389
20. Toroslu I H, Ücoluk G (2007) Incremental assignment problem. Inform Sci 177:1523–1529
21. Nagarajan R, Solairaju A (2010) Computing Improved Fuzzy Optimal Hungarian Assignment Problems with Fuzzy Costs under Robust Ranking Techniques. Int J Comput Appl 6:6–13
22. Ayorkor G, Mills-Tettey, Anthony Stentz, Bernardine Dias M (2007) The Dynamic Hungarian Algorithm for the Assignment Problem with Changing Costs. Robotics Institute. Paper 149
23. Zhu H, Zhou M C, Alkins R (2012) Group role assignment via a Kuhn–Munkres algorithm-based solution. IEEE Trans Syst Man Cybern, Part A, Syst Humans 42:739–750
24. Zhu H, Liu D, Zhang S, Zhu Y, Teng L, Teng S (2016) Solving the many to many assignment problem by improving the Kuhn–Munkres algorithm with backtracking. Theor Comput Sci 618:30–41
25. Zhang X, Zhang J, Gong Y, Zhan Z, Chen W, Li Y (2016) Kuhn–Munkres parallel genetic algorithm for the set cover problem and its application to large-scale wireless sensor networks. IEEE Trans Evolut Comput 20:695–710
26. Laha D, Gupta J N (2016) A Hungarian penalty-based construction algorithm to minimize makespan and total flow time in no-wait flow shops. Comput Ind Eng 98:373–383
27. Hamuda E, Mc Ginley B, Glavin M, Jones E (2018) Improved image processing-based crop detection using Kalman filtering and the Hungarian algorithm. Comput Electron Agr 148:37–44
28. Jamal J, Shobaki G, Papapanagiotou V, Gambardella L M, Montemanni R (2017) Solving the sequential ordering problem using branch and bound. In: Proceedings of IEEE SSCI 2017 3110–3118.
29. Singh S, Dubey G C, Shrivastava R, A Comparative Analysis of Assignment Problem IOSR J Eng 2:01–15
30. Schäfer F, Walther M, Hübner A, Kuhn H (2019) Operational patient-bed assignment problem in large hospital settings including overflow and uncertainty management. Flexible Services and Manufacturing Journal, 1–30.

31. Vancroonenburg W, De Causmaecker P, Vanden Berghe G (2016) A study of decision support models for online patient to-room assignment planning. Annals of Operations Research 239(1):253–271.
32. Taramasco C, Olivares R, Munoz R,Soto R , Villar M, Hugo V (2019) The patient bed assignment problem solved by autonomous bat algorithm. Applied Soft Computing Journal 81 105484.
33. Zhu Y, Toffolo T A M, Vancroonenburg W, Vanden Berghe G (2018) Compatibility of short and long term objectives for dynamic patient admission scheduling, Computers and Operations Research, https://doi.org/10.1016/j.cor.2018.12.001

Chapter 3
A Bi-objective Algorithm for Robust Operating Theatre Scheduling

Tarek Chaari and Imen Omezine

Abstract Operating theatre is considered as a critical activity in health care system. Operating theatre managers must be taken into consideration uncertainties of treatment durations. However, this is a challenging task because it is almost impossible to predict the exact duration of a surgery. In order to help operating theatre managers build robust schedules, we develop in this study a robust bi-objective algorithm, which takes into account uncertainty of surgical and recovery durations. Besides, we integrate uncertainty in terms of a scenario-based optimization approach. A robust bi-objective evaluation function was defined to obtain a robust, effective solution that is only slightly sensitive to data uncertainty. These bi-objective functions minimize the Makespan of the initial scenario, and the deviation between the Makespan of all the disrupted scenarios and the Makespan of the initial scenario. We validated our method with a simulation to evaluate the quality of the robustness faced with uncertainty. The computational results show that our algorithm can generate a trade-off for effectiveness and robustness for various degrees of uncertainty.

3.1 Introduction

The managerial aspect of providing health services to patients in hospitals is becoming increasingly important. On the one hand, hospitals want to reduce their costs and improve their financial assets and, on the other hand, maximize the level of patient satisfaction.Knowing that the surgery department usually accounts for around 10% of a hospital budget[10], one can understand the importance of an optimal usage of operating theatre (OT). However, this is a challenging task because it is almost impossible to predict the exact duration of a surgery. In order to help OT managers build robust schedules, we develop in this study a tool, which takes into

T. Chaari (✉) and I. Omezine
Management Information System Department, Taibah University, Medina, Kingdom of Saudi Arabia
e-mail: tchaari@taibahu.edu.sa; iomezine@taibahu.edu.sa

© Springer Nature Switzerland AG 2021
M. Masmoudi et al. (eds.), *Operations Research and Simulation in Healthcare*, https://doi.org/10.1007/978-3-030-45223-0_3

account uncertainty to obtain a robust and effective solution that is only slightly sensitive to data uncertainty. However, in reality, several sources of uncertainty can have an effect on an operating theatre plan, and the managers are unable to provide reliable or satisfactory data for the problems and delay time that arise. The presence of uncertainty makes the question of modeling it inevitable. The literature offers many modeling approaches that vary in terms of their formalism as well as their processing mode.

The objective of this work is to develop a Multi-Objective Genetic Algorithm (MOGA) to obtain a robust and effective solution that is only slightly sensitive to data uncertainty. We choose the scenario modeling to represent the uncertainty of the problem characteristics.

The work is organized as follows: We present in Sect. 3.2 three modeling approaches: stochastic and probabilistic modeling, fuzzy modeling and scenario modeling. Section 3.3 provides a review of relevant literature, and Sect. 3.4 presents a bi-objective algorithm for robust scheduling. Section 3.5 gives the test results, and we conclude and outline directions for further research in Sect. 3.6.

3.2   Uncertainty Modeling

Generally, the knowledge of real systems is imperfect. According to [1], this imperfect knowledge is exemplified by uncertainty, imprecision and/or incompleteness. The concepts of uncertainty, imprecision and incompleteness are interrelated. In fact, incompleteness implies uncertainty. Imprecision can also be associated with incompleteness, thus generating uncertainty. For this reason, we refer to uncertainty, which is the most global concept. Many criteria have been proposed to classify the various types of uncertainty. For example, Bonfill [2] presented taxonomy of uncertainty, in which he differentiated between strategic, tactical and operational uncertainties.

The duration of a case is hard to predict because it can vary widely from one case to another, even for identical procedures. The surgeon performing the surgery, age and risks associated with the patient condition, emergency, type of anaesthesia and gender are all responsible for part of the uncertainty. Abedini et al. [15] take both sources of uncertainties into consideration: uncertainties in the case times and patients arrivals at the same time. These authors present two types of uncertainties in OT planning. Uncertainty in the case times [12, 13], which is the difference between expected and actual surgery duration and uncertainty in the patients arrivals caused by emergency arrivals and patients no-show cases [11] and [14].

Modeling uncertainty in scheduling implies a compromise between a rich representation that is close to reality, and one that is clear and understandable [3]. The modeling process forces designers to make valid arbitrary choices to simplify and aggregate. French [4] noted that the theoretical modeling choices are generally carried out implicitly and are largely influenced by the training and aptitudes of the designer and by the tools that he or she is used to handling. The literature offers many modeling approaches that vary in terms of their formalism as well as their processing mode. We present three modeling approaches.

3.2.1 Stochastic and Probabilistic Modeling

Probabilistic modeling implies identifying the probability distribution in which the problem uncertainties are represented by random variables. However, determining the appropriate probability distribution precisely from the theoretical distributions is difficult. This difficulty causes ambiguity, which can be interpreted as due to the imprecision of the probabilities. This ambiguity provokes a situation that is midway between a risky situation in the strictest sense (i.e. only one distribution, whose parameters are known) and a fundamentally uncertain situation (i.e. several distributions are possible without being able to determine which one is the most suitable) [5]. For more details, see [15, 21] and [19].

3.2.2 Fuzzy Modeling

Uncertainties are not always of a random nature. Fuzzy modeling, based on the fuzzy subsets theory, is the preferred approach for modeling imprecise situations [6]. Fuzzy sets are used in context where data are affected by uncertainty and/or vagueness [7]. This approach allows the representation of classes whose limits are poorly defined. These variations are represented by the membership functions. Membership is gradual, which allows imprecise knowledge (e.g. the information collected in natural language, or approximate values due to measurement difficulties) to be expressed.

3.2.3 Scenario Modeling

The scenario modeling approach requires the construction of a set of scenarios (also called instances or data files) containing numerical values that reflect the hypotheses about the uncertainty. The uncertainty can then be modeled using discrete or continuous scenario sets (through intervals with known and certain terminals). Reference [8] used a set of scenarios involving stochastic inputs. Scenario-based approaches are widely used for robust scheduling. Li et al. [9] proposed a multi-period scenario-based programming model in which all the decision variables are treated as first-stage variables in a two-stage stochastic programming model. The interested reader can consult [24, 29] and [30] for more details about this modeling approach.

3.3   Literature Review

In the past 50 years, a large body of literature on the operating theatre scheduling has evolved. One of the major and difficult problems associated with the accurate surgery scheduling is the uncertainty inherent to surgical services, such as surgery durations.

Abedini et al. [15] develop a stochastic bi-level optimization models were formulated to minimizing total cost and maximizing throughput of operating rooms under uncertainties in patient arrivals and duration times of surgery. The authors propose a simulation-based trade-off balancing model to minimize the sum of deviations from best performance on each objective. Using historical data obtained from real case.

Bruni et al. [21] use a stochastic model with recourse for the scheduling of surgeries when the occurrence of emergency patients and surgery durations are uncertain. Three recourse strategies are introduced to be taken on a daily basis, comprising : the overtime recourse strategy, the swapping recourse strategy and the complete rescheduling strategy.

Liu et al. [26] solve an operating room scheduling and surgeon assignment problem with uncertain surgery durations. They authors develop a stochastic model under the real-life constraints. The objectives are to increase the utilization and reduce the cost of operating theatre and to improve surgeons' satisfaction in the meantime. These authors consider surgery duration uncertainty and propose a two-step mixed integer programming model. Saadouli et al.[27] presented a stochastic model for the scheduling problem under durations and postoperative recovery time uncertainty. Dexter [16] proposes a statistical method to predict whether moving the last surgery of the day to another OR would reduce the overtime labor costs.

Wachtel and Dexter [20] present several interventions including rescheduling to reduce tardiness in surgical suites. Stuart et al.[18] study a single-OR rescheduling problem by treating reactive scheduling as an assignment problem with the objective of minimizing the cancellation of elective surgeries, minimizing deviation from the original schedule and maximizing the throughput of emergent surgeries. Nouaouri et al. [23] propose a 3-stage integer programming model for reactive scheduling in the case of a disaster in order to optimize the operating rooms scheduling taking into account disruptions with insertion of unexpected new patient in the pre-established operating schedule. Stuart and Kozan [19] present an innovative reactive scheduling model that was developed and solved for a single operating theatre. Two types of disruptions are defined for this problem: an operating theatre disruption and patient disruption. The model is run in real time following the completion of each operation and solved with a branch-and-bound heuristic. The criterion was to maximize the weighted number of expected on-time patients.

Hans et al. [22] propose a robust scheduling to deal with the under-estimation of surgical durations caused by their inherent variability. In their paper, a statistically based amount of slack or buffer is added to the amount of time assigned for each surgery to absorb the effects of variations in treatment time that occur in the online setting. Surgical duration estimates are based on historical data, and the planned slack time is based on the volatility of the surgical duration estimates. Geranmayeh[24] aims at minimizing the peak load in the surgical ward by considering uncertainty in patient length of stay. Her model includes operating room and surgeon availability. Furthermore she utilizes robust optimization to protect against the uncertainty in the number of inpatients a surgeon may send to the ward per block. Different scenarios were developed that explore the impact of varying the availability of operating rooms on each day of the week. The models were verified by simulation model. Rachuba and Werners[28] present a robust compromise approach in order to support surgery scheduling in the presence of multiple stakeholders and uncertain demand. The robust extension to the compromise approach was based on a fuzzy sets approach that adequately models the differing interests. Different schedules are calculated and later on evaluated with randomly generated scenarios for surgery times and emergencies.

Dexter et al. [17] used a simulation to evaluate the relative performance of algorithms described in the literature for scheduling add-on elective cases into open operating room time. Semi-urgent surgeries that should be performed soon but not necessarily today on a shared operating room are examined by [25]. They authors develop a methodology to handle the semi-urgent patient flow at a surgical department.

The scheduling problems, in the industrial environment, have been largely addressed. The literature, which is abundant in this area, is inspired by proven tools in the industry. In the hospital context, scheduling problems within operating rooms have often been assimilated to hybrid flow shops with two stages [31–33]. The number of stages depends on the number of resources considered. The stages represent operating rooms and recovery rooms that are parallel and identical. Each patient must be treated first in an operating room and next in a recovery room without waiting time. The operating theatre is composed of several operating rooms and several beds that are available in the recovery room. The patient is blocked in operating room, if no bed is available in recovery room. The job processing times are equal to intervention duration in the first stage and to recovery duration in the second stage.

In the industrial environment, many researchers are used a technique based on robustness measure to obtain a robust solution (scheduling). For example, Kouvelis and Yu [38] develop 3 measures of robustness: The *absolute robustness* of a decision is defined as the worst-case performance of a certain decision across all scenarios. The *robust deviation* of a decision is the deviation of the performance of this decision from the performance of the best possible decision for each scenario. Finally, the *relative robustness* is the worst observed percentage deviation from optimality for the evaluated decision across all scenarios. Sevaux and Sorensen [39] develop a genetic algorithm for a single machine scheduling problem in order to find a robust solution. A possible form of a robust evaluation function is a *weighted average* of different derived evaluations. For more details about these measures of robustness, the interested reader can consult [36].

The major disadvantage of all the robustness techniques mentioned above is that the solution obtained does not give enough flexibility to decision makers to choose the best robust solution for their various objectives. To overcome this disadvantage, we propose a bi-objective algorithm for robust and effective scheduling in the following part.

3.4 A Bi-objective Algorithm for Robust Scheduling

3.4.1 Problem Definition

In this work, a robust bi-objective evaluation function is used. It allows minimizing simultaneously two objectives. The first objective minimizes the value of the criterion of the initial scenario. The second objective minimizes the deviation between the value of the criterion of all disrupted scenarios and the one of the initial scenario. The goal of this evaluation function is to obtain an effective solution that is not very sensitive to data uncertainty.

The transformation of the multi-objective problem into a single objective problem requires a priori knowledge of the problem addressed. The optimization of a single objective problem can guarantee the optimality of the solution found but find a single solution. In real-life situations, decision makers generally need several alternatives, if certain objectives are noisy or data are uncertain, these methods are not effective. The notion of dominance is necessary for this type of problem.

Many of the real-world problems consider simultaneous optimization of several objectives. These objectives are usually in competition and conflict with each other. Multi-objective optimization allows to provide a solution set using the conflict between objectives that is called pareto-optimal. This solution set includes solutions that no other solution could have found when all objectives were taken into account.

$$Min\ f(x) = (f_1(x), f_2(x), \ldots, f_k(x)), \tag{3.1}$$

$$\text{where } X \in R^n$$

where the integer $k \geq 2$ is the number of objectives, and the set X is the feasible set of decision vectors.

A solution x_1 is said to dominate another solution x_2 if:

1. $f_i(x_1) \leq f_i(x_2)$ for all indices $i \in \{1, 2, \ldots, k\}$,
2. $f_j(x_1) < f_j(x_2)$ for at least one index $j \in \{1, 2, \ldots, k\}$.

Solutions that dominate others but do not dominate themselves are called non-dominated solutions. According to their definition, vector x_1 is a globally pareto-optimal solution if vector x_2 does not exist such that x_2 dominates x_1. The set of Pareto optimal outcomes is often called the Pareto frontier.

To solve our multi-objective problem, we used the Multi-Objective Genetic Algorithm (MOGA) proposed by [34]. This algorithm is based on the relationship of dominance that aims to diversify the results and increase their performance. The algorithm uses a fitness function to take into account the rank of the individual and the number of individuals with the same rank. That is, all non-dominated solutions in the population are ranked 1, and solutions dominated by one or more points are ranked 2 or higher.

3.4.2   MOGA for Robust Operating Theatre Scheduling

In this section, we develop a MOGA for robust operating theatre scheduling to obtain non-dominated solutions. This algorithm aims to generate a robust, good-quality solution in terms of the defined criterion. To obtain a robust solution, the fitness evaluation function is replaced by a robust evaluation function.

3.4.2.1   Uncertainty Modeling

To represent the uncertainty about the characteristics of the problem, we chose a scenario modeling approach that requires constructing a set of scenarios containing the numerical values that reflect the hypotheses about uncertainty.

In this study, the surgery durations (in the first stage) and recovery duration (in the second stage) are considered to be uncertain. I is an initial scenario, which represents the characteristics (i.e. inputs) of the problem. The set of the scenarios, denoted by a function ξ, is represented with a sampling function that starts with the initial scenario I.

Let $\xi_i(I)$ be the ith set of the sampled parameters from the initial scenario I (where $i = 1, 2, \ldots, N$ and N represents the number of disrupted scenarios), and note that $\xi_i(I)$ are uniformly distributed between $[P_I - \alpha P_I; P_I + \alpha P_I]$, where is P_I the processing time of the initial scenario and $\alpha \in [0, 1]$ is used to express the degree of uncertainty of the processing times. The uniform is a common distribution [35] and used to experiment our approach.

3.4.2.2   Robustness Criteria

The evaluation of the robust solution is based on a bi-objective robustness criterion. Let x be a solution to the problem (i.e. a permutation of the jobs). The quality of x is computed by a robust evaluation function $f_r(x)$ (or robustness criterion), where the criterion is the completion date of the last patient (makespan). The robust evaluation function is defined as follows:

$$f_r(x) = (f_1(x), f_2(x)) , \qquad (3.2)$$

where:

$$f_1(x) = Cmax_I(x),$$

$$f_2(x) = \sqrt{\frac{1}{N} \sum_{i=1}^{N} \left(Cmax_{\xi_i(I)}(x) - Cmax_I(x)\right)^2},$$

$Cmax_I(x)$ is the makespan of initial scenario I,
$Cmax_{\xi_i(I)}(x)$ is the makespan of disrupted scenario $\xi_i(I)$.

This bi-objective function minimizes simultaneously the makespan of the initial scenario and the deviation between the makespan of all disrupted scenarios and the makespan of the initial scenario. We note that these two objectives $f_1(x)$ and $f_2(x)$ have been proposed by [37]; however, these authors have used a parameter λ varied in 0 and 1 to aggregate these two objectives on a single objective. In the MOGA we used two functions separately to obtain a pareto set solutions.

3.4.2.3 Description of Our MOGA for Robust Operating Theatre Scheduling

1. *Coding scheme*: Each chromosome represents a schedule directly [40], a direct coding approach was used. We chose to maintain a simple permutation of n patients (jobs) indicating the order in which the patients are processed in the operating rooms at the first stage. For the second stages (recovery rooms), the patients are ordered according to their minimal completion time of the preceding stage, i.e. the queue of patients in the second stages is processed by the First In First Out (FIFO) rule. The patients are assigned to operating rooms and recovery rooms by applying the first available machine (FAM) assignment rule.
2. *Initial population*: The initial population is randomly generated. This population P is composed of a given number of chromosomes.
3. *Evaluation*: The value of each objective function is calculated for every chromosome and assign to each chromosome a rank based on the dominance rule. All non-dominated individuals are assigned rank 1 and any other individual, $\tilde{x}_i$, is assigned a rank equal to the number of solutions, n_i, that dominate solution $\tilde{x}_i$ plus one. Thus, for solution $\tilde{x}_i$ the rank assigned is $r_i = 1 + n_i$.
4. *Parent selection*: The method used is the selection of a pair of chromosome (parents) randomly, one parent has a rank 1 and other parent has any other rank different from 1.
5. *Crossover operator*: The Two-Point Order Crossover (TP) operator is used in our method. This operator is defined as follows: two points are randomly selected for dividing one parent. The elements outside the selected two points are always inherited from the parent to the child. The elements inside will be transferred from the second parent as they appear from left to right. Two children are produced in this way.

6. **Mutation operator**: A shift mutation operator is used, which randomly selects a gene from the chromosome and inserts it at another random position in the chromosome.

7. **Formation of child population**: After generating a full set of offspring, the parent population is combined with this set of offspring to form a new population (equal to $2P$).

8. **Insertion**: The new population obtained after mutation is evaluated and ranked. The set of Pareto optimal solutions is updated and only the higher half of the new population, corresponding to the best chromosomes, is selected for the next iteration. Thus the size of the population remains constant (equal to P) from generation to generation.

9. **Stopping criterion**: The stopping criterion is a fixed number of generations (or iterations) equal to $Iter_{Max}$. If the terminating condition is satisfied, the set of non-dominated solutions of the final generation is presented to the decision maker. The best solution is then selected according to the decision maker's preference.

Program Code: MOGA for Robust Scheduling

```
 1 : Initialize N, P, P_cros, P_mut, t=0, Iter_max,
 2 : Generate N disrupted scenario.
 3 : Generate randomly P initial Population.
 4 : Calculate all objective functions for each solution in P.
 5 : Specify rank for each solution in P.
 6 : Repeat
 7 :    i=0
 8 :         While t<P/2,
 9 :             Select randomly two parents from population P.
                 one of rank 1 and one of any rank.
10:             Apply the crossover scheme based on P_cros .
11:             Apply the mutation scheme based on P_mut.
12:             t=t+1
13:         End While
14:     Calculate all objective functions for new population.
15:     The newly generated offspring and the parent population
        equal 2P, are combined together and ranked.
16:     Select the higher half of the best chromosomes.
17:     i=i+1
18: Until  i=Iter_max,
19: Identify the best solutions with rank=1 as the final
    non-dominated Pareto sets.
```

3.5   Computational Experiments

Our MOGA is implemented in C++ on an Intel(R) CORE(TM) i7 with 2.4GHz and 4 GB of RAM. We conducted several experimental tests to choose the best parameter values for our method. We retained the following values:

- Population size $P = 100$;
- Degree of uncertainty $\alpha = 20\%$;
- Number of disrupted scenario $N=20$;
- Crossover probability $P_{cross} = 0.9$;
- Mutation probability $P_{mut} = 0.1$;
- Stopping criterion $Iter_{Max} = 1000$ generations.

3.5.1   The Test Problem

The performance of our method for the operating theatre scheduling problem was examined by solving 20 different problems involving 4 different sizes with the numbers of patient interventions are (10, 15, 20, 30). The number of operating rooms is (2, 3, 5), and the number of beds in recovery room is (3, 5, 8). Based on the data obtained from [41], the surgery durations are classified into four types: small (S), medium (M), large (L) and extra-large (EL). The durations of these five surgery types and the recovery duration are represented as Normal distribution (μ, σ) and are listed in Table 3.1. Note that the surgery duration includes the duration of the surgical procedure, preparation, anaesthesia and clean-up time.

3.5.2   Computational Results

We validated the MOGA using simulation to highlight and measure its effectiveness at generating robust solutions. We used the ARENA 16.0 simulation software to create a simulation model, which allowed us to test and validate the robustness obtained by MOGA. The simulation model corresponds to a possible organization of the operating theatre with 2 stages. The first stage represents the identical operating rooms, and the second stage represents the recovery room comprised of several identical beds. The MOGA allows a robust sequence to be calculated,

Table 3.1 The different types of surgery durations and recovery duration

	Surgery durations				Recovery duration
	S	M	L	EL	
Duration	$N(33, 15)$	$N(86, 17)$	$N(153, 17)$	$N(213, 17)$	$N(28, 17)$

which minimizes the robust evaluation function. As input, the simulation receives the sequence of the patients obtained by MOGA. We introduced stochastic data into every simulated scenario in the model. These data were randomly generated according to the uniform distribution between $[P_I - \alpha P_I; P_I + \alpha P_I]$, such as defined in Sect. 3.4.2.1. The assignment rules for the patients in the operating theatre are the same as the assignment rules used in the MOGA.

The MOGA was run to obtain a sequence. Once this sequence was obtained, the simulator received this sequence of the patients as input. We evaluated 50 replications of the problem instance using this sequence and based on stochastic data.

Tables 3.2, 3.3, 3.4 and 3.5 give the detailed results for the MOGA for the set of instances with a degree of uncertainty equal to 20%. In these tables, the

Table 3.2 Experimental results with 10 patients

Instance	LB	$f_1(x)$	$f_2(x)$	$\overline{X}_{simulation}$	%Deviation in %
J10-1	405	405	26.59	416.42	**2.82**
		407	25.79	415.71	**2.14**
		412	25.49	419.70	**1.87**
		413	23.66	419.57	**1.59**
		420	22.82	425.92	**1.41**
		422	22.80	425.63	**0.86**
		436	22.23	438.49	**0.57**
		449	21.64	449.54	**0.12**
J10-2	675	675	33.24	693.63	**2.76**
		680	29.96	698.09	**2.66**
		681	29.48	695.10	**2.07**
		683	27.13	694.61	**1.70**
		687	26.85	696.27	**1.35**
		690	22.95	697.14	**1.03**
		699	20.47	697.67	**−0.19**
J10-3	594	594	28.22	607.42	**2.26**
		597	27.57	610.37	**2.24**
		598	26.24	609.48	**1.92**
		606	25.62	614.91	**1.47**
		610	23.96	614.09	**0.67**
		616	21.89	618.40	**0.39**
J10-4	805	806	28.25	821.14	**1.87**
		808	27.45	822.85	**1.83**
		809	26.51	817.47	**1.04**
		812	24.65	818.76	**0.83**
J10-5	704	704	23.76	719.52	**2.20**
		712	25.25	722.85	**1.52**
		718	29.15	726.76	**1.22**
		734	34.61	732.33	**−0.22**

Table 3.3 Experimental results with 15 patients

Instance	LB	$f_1(x)$	$f_2(x)$	$\overline{X}_{simulation}$	%Deviation in %
J15-1	448	450	20.10	462.19	**2.70**
		453	18.94	464.14	**2.45**
		454	16.02	459.38	**1.18**
		464	13.18	466.85	**0.61**
		465	12.23	467.14	**0.46**
J15-2	207	215	8.79	221.57	**3.05**
		219	7.38	225.42	**2.93**
		224	5.82	228.76	**2.12**
		231	5.49	235.19	**1.81**
		232	5.47	234.44	**1.05**
		242	5.32	241.85	**−0.06**
J15-3	821	832	32.12	870.23	**4.59**
		833	27.49	859.90	**3.22**
		836	24.85	861.33	**3.03**
		839	24.65	860.14	**2.52**
		842	19.14	861.37	**2.30**
		844	18.92	858.01	**1.66**
		849	17.15	858.34	**1.10**
		855	16.12	861.84	**0.80**
J15-4	206	210	12.80	221.52	**5.48**
		212	8.769	219.19	**3.39**
		214	7.234	220.19	**2.89**
		216	7.044	220.28	**1.98**
		217	6.395	220.09	**1.42**
		220	6.102	222.73	**1.24**
		221	6.106	223.41	**1.09**
J15-5	1110	1110	44.71	1141.52	**2.84**
		1111	44.46	1133.14	**1.99**
		1112	40.21	1132.91	**1.88**
		1113	36.85	1132.70	**1.77**
		1114	34.36	1132.61	**1.67**
		1116	34.06	1131.71	**1.40**
		1120	29.84	1137.47	**1.56**
		1124	27.37	1131.14	**0.63**
		1126	26.93	1132.90	**0.61**
		1130	23.23	1130.90	**0.08**

Table 3.4 Experimental results with 20 patients

Instance	LB	$f_1(x)$	$f_2(x)$	$\overline{X}_{simulation}$	%Deviation in %
J20-1	564	564	22.75	578.85	2.63
		567	16.10	579.36	2.18
		568	15.62	579.33	1.99
		573	14.37	584.06	1.93
		575	14.00	584.89	1.72
		576	11.99	585.22	1.60
		581	11.34	588.08	1.21
		594	11.01	591.71	−0.38
J20-2	1449	1471	51.31	1497.57	1.80
		1488	48.55	1513.95	1.74
		1493	45.48	1518.52	1.70
		1502	40.49	1511.28	0.61
		1533	40.25	1539.29	0.41
		1553	37.94	1556.90	0.25
J20-3	614	622	23.98	640.80	3.02
		623	17.01	634.14	1.78
		628	16.46	638.04	1.59
		631	15.27	640.65	1.52
		634	15.04	643.14	1.44
		638	14.75	646.61	1.35
		639	14.61	647.14	1.27
		642	14.45	645.38	0.52
		658	14.38	660.76	0.41
J20-4	1005	1005	34.04	1023.49	1.84
		1023	26.62	1039.78	1.64
		1028	26.61	1039.52	1.12
		1035	23.67	1044.42	0.91
		1061	23.61	1062.47	0.13
J20-5	602	604	14.47	621.52	2.90
		605	13.37	617.33	2.03
		611	12.10	618.23	1.18
		613	10.41	619.33	1.03
		621	10.01	625.97	0.80
		623	9.17	626.71	0.59

Table 3.5 Experimental results with 30 patients

Instance	LB	$f_1(x)$	$f_2(x)$	$\overline{X}_{simulation}$	%Deviation in %
J30-1	936	954	37.34	976.19	**2.32**
		955	33.65	974.57	**2.04**
		956	32.12	971.20	**1.59**
		963	27.60	977.64	**1.52**
		973	24.70	983.51	**1.08**
		975	24.41	984.26	**0.95**
		978	21.59	984.94	**0.71**
J30-2	927	934	30.85	965.66	**3.39**
		944	27.99	967.38	**2.47**
		949	25.70	966.42	**1.83**
		958	23.97	975.04	**1.77**
		963	22.28	976.66	**1.41**
		967	21.66	974.52	**0.77**
		969	21.47	975.40	**0.66**
		974	19.58	977.80	**0.39**
		995	19.36	982.04	**−1.30**
J30-3	1317	1327	40.97	1372.91	**3.46**
		1337	38.16	1378.05	**3.07**
		1339	37.78	1378.57	**2.95**
		1340	37.58	1377.79	**2.82**
		1341	35.83	1376.57	**2.65**
		1345	35.47	1373.78	**2.14**
		1346	34.13	1372.92	**2.00**
		1347	30.98	1371.23	**1.79**
		1353	26.83	1374.92	**1.62**
		1355	24.67	1376.82	**1.61**
J30-4	200	204	10.24	210.80	**3.33**
		205	9.68	211.09	**2.97**
		206	8.66	211.71	**2.77**
		207	8.51	211.95	**2.39**
		208	6.86	211.19	**1.53**
		211	5.18	210.85	**−0.07**
J30-5	1315	1326	49.39	1385.40	**4.48**
		1331	48.40	1384.42	**4.01**
		1332	45.06	1383.14	**3.83**
		1334	36.84	1382.29	**3.62**
		1336	31.03	1377.19	**3.08**
		1341	29.22	1378.41	**2.79**
		1342	23.66	1377.19	**2.62**
		1350	21.26	1376.42	**1.95**
		1358	18.76	1376.74	**1.38**

column 'Instance' gives the number of problem instance. The column 'LB' gives the lower bound of the initial scenario (note that we use the lower bound developed by Santos et al. [42] for the test problems randomly generated). The column $f_1(x)$ (the first objective) gives the non-dominated solutions of the makespan value of the initial scenario obtained by MOGA. The column $f_2(x)$ (the second objective) is the deviation between the makespan of all disrupted scenarios and the makespan of the initial scenario. The column $\bar{X}_{simulation}$ presents the average of value for 50 replications. The column '% Deviation' gives the deviation or the $Cmax$ increase (in %) of the 50 disrupted scenarios compared to the initial scenario's makespan, where $\% Deviation = \frac{\bar{X}_{simulation} - Cmax_I}{Cmax_I}$.

According to Tables 3.2, 3.3, 3.4 and 3.5, the more the makespan of the initial scenario increases, the more the deviation between the makespan of all disrupted scenarios and the makespan of the initial scenario decreases.

Let us examine in detail the instance j10-1. We found 8 non-dominated solutions. The low makespan of the initial scenario obtained is 405, which is close (or equal) to the LB. If we maintain this sequence, knowing that processing time can be disrupted with degree of uncertainty equal to 20%, we could expect an average makespan of 416.42 after the simulation, which is a cost increase of 2.82%. Therefore, this solution is sensitive to uncertainty but gives good results for Cmax.

If we take the same instance, the smaller value of $f_2(x)$ is equal to 21.64 and gives a value of Cmax=449. If this sequence is used in the disrupted scenarios, the expected makespan value of the initial scenario will be an average of 449.54, which is an increase of only 0.12%. This solution is then more robust in terms of managing uncertainty but not effective in terms of makespan.

According to Tables 3.2, 3.3, 3.4 and 3.5, the '% Deviation' decreases when $Cmax$ increases. This decrease means that more confidence should be granted to the robust solution even if the makespan of the initial scenario is higher.

3.6   Conclusion

In this paper, we proposed a multi-objective genetic algorithm for robust operating theatre scheduling problem with uncertain durations. The objective is to minimize simultaneously the makespan of the initial scenario to obtain an effective solution and the deviation between the makespan of all disrupted scenarios and the makespan of the initial scenario in order to obtain a robust solution. Our algorithm generates non-dominated solutions, these solutions have been validated by simulation. The simulation results show that the proposed algorithm can generate a trade-off between effectiveness and robustness. Our MOGA can be used towards a decision-support system since it already helps operating theatre managers by providing an off-line schedule that respects the overall traditional operational constraints.

From a general methodological perspective, the algorithm proposed in this work may be further extended to interesting avenues of research. One such avenue would be to apply in the first step a robust scheduling and the second step a reactive scheduling (on-line scheduling) is applied to integrating the patients arrivals in emergency.

References

1. Bouchon-Meunier, B., 1995: La logique floue et ses applications. Addison-Weyley.
2. Bonfill, A. 2006. Proactive management of uncertainty to improve scheduling robustness in process industries. thesis (Ph.D). Universitat Politecnica de Catalunya.
3. Bouyssou, D., 1989: Modelling inaccurate determination, imprecision using multiple criteria. In Lockett, A.G. et Islei, G. (Eds.), Improving Decision Making in Organisations (page 78–87). Springer- Verlag, Heidelberg.
4. French, S., 1995: Uncertainty and imprecision Modelling and analysis. Journal of the operational research society, 46, 70–79.
5. Einhorn, H.J., Hogarth, R.M., 1985: Ambiguity and uncertainty in probabilistic inference. Psychological Review, 92(4), 433–461.
6. Zadeh, L.A., 1965: Fuzzy sets, Information and Control, 8, 338–353.
7. Cigolini, R., Rossi, T., 2008: Evaluating supply chain integration: a case study using fuzzy logic. Production Planning and Control, 19(3), 242–255.
8. Leung, S., Yue, W., 2004: A robust optimization model for stochastic aggregate production planning. Production Planning and Control, 15(5), 502–514.
9. Li, I., Karimi, A., Srinivasan, R., 2006: Robust Scheduling of Crude Oil Operations under Demand and Ship Arrival Uncertainty. The 2006 Annual Meeting San Francisco, CA, Systems and Process Design (10a), Computing and Systems Technology Division.
10. Chaabane, S., Guinet, A., Smolski, N., Guiraud, M., Luquet, B., Marcon, E. Viale, J.-P, 2003: La gestion industrielle et la gestion des blocs opératoires. Annales Francaises d'Anesthésie et de Réanimation 22(10), 904–908.
11. Cardoen, B., Demeulemeester, E., Beliën, J., 2010: Operating room planning and scheduling: A literature review. European journal of operational research, 201(3), 921–932.
12. Lamiri, M., Xie, X., Dolgui, A., Grimaud, F., 2008: A stochastic model for operating room planning with elective and emergency demand for surgery. European Journal of Operational Research, 185(3), 1026–1037.
13. Denton, B.T., Miller, A.J., Balasubramanian, H.J., Huschka, T.R., 2010: Optimal allocation of surgery blocks to operating rooms under uncertainty. Operations research, 58(4-part-1), 802–816.
14. Epstein, R.H., Dexter, F., 2013 :Rescheduling of previously cancelled surgical cases does not increase variability in operating room workload when cases are scheduled based on maximizing efficiency of use of operating room time. Anesthesia & Analgesia, 117(4),995–1002.
15. Abedini, A., Li, W., Ye,H., 2018 :Stochastic bi-level optimization models for efficient operating room planning. Procedia Manufacturing, 26, 58–69.
16. Dexter, F., 2000: A strategy to decide whether to move the last case of the day in an operating room to another empty operating room to decrease overtime labor costs. Anesthesia & Analgesia, 91(4), 925–928.
17. Dexter, F., Macario, A., Traub, RD., 1999 : Which algorithm for scheduling add-on elective cases maximises operating room utilization Use of bin packing algorithms and fuzzy constraints in operating room management. Anesthesiology 91(5):1491–1500.
18. Stuart, K., Kozan, E., Sinnott, M., Collier,J., 2010: An innovative robust reactive surgery assignment model. ASOR Bulletin, 29(3):48–58.

19. Stuart, K., Kozan, E., 2012: Reactive scheduling model for the operating theatre. Flexible Services and Manufacturing Journal. 24, 400–421.
20. Wachtel, R.E., Dexter, F., 2009: Reducing tardiness from scheduled start times by making adjustments to the operating room schedule. Anesthesia & Analgesia, 108(6):1902–1909.
21. Bruni, M., Beraldi, P., Conforti, D., 2015: A stochastic programming approach for operating theatre scheduling under uncertainty. IMA Journal of Management Mathematics, 26(1):99–119.
22. Hans, E., Wullink, G., Van Houdenhoven, M., Kazemier, G., 2008: Robust surgery loading. European Journal of Operational Research. 185(3):1038–1050.
23. Nouaouri, I., Nicolas, J-C., Jolly, D., 2010 : Reactive operating schedule in case of a disaster: arrival of unexpected victims. Proceedings of the world congress on engineering, vol III. London, UK.
24. Geranmayeh, S., 2015: Optimizing surgical scheduling through integer programming and robust optimization. Master thesis, university of Ottawa.
25. Zonderland, M.E., Boucherie, R.J., Litvak, N., 2010: Planning and scheduling of semi-urgent surgeries. Health Care Management Sciences. 13: 256–267.
26. Liu, H., Zhang, T., Luo, S. , Dan Xu, D., 2018 : Operating room scheduling and surgeon assignment problem under surgery durations uncertainty.Technology and Health Care 26, 297–304.
27. Saadouli, H., Jerbi, B., Dammak, A.,2015 : A stochastic optimization and simulation approach for scheduling operating rooms and recovery beds in an orthopedic surgery department. Computers and Industrial Engineering, 80, 72–79.
28. Rachuba, S., Werners, B., 2014: A robust approach for scheduling in hospitals using multiple objectives. Journal of the Operational Research Society, 65, 546–556.
29. Liu, C., Wang, J.J., Liu,M., 2017: A scenario-based robust optimization approach for surgeries scheduling with a single specialised human resource server. International Conference on Service Systems and Service Management Service Systems and Service Management (ICSSSM):1–4 Jun, 2017.
30. Heydari, M., Soudi, A., 2016: Predictive Reactive Planning and Scheduling of a Surgical Suite with Emergency Patient Arrival. Journal of Medical Systems. 40, 30.
31. Guinet, A., Chaabane, S., 2003: Operating theatre planning, International Journal of Production Economics, 85(1), 69–81.
32. Fei, H., Meskens, N., Chu, C., 2010 : A planning and scheduling problem for an operating theatre using an open scheduling strategy', Computers & Industrial Engineering, 58(2),221–230.
33. Jebali, A., Hadj Alouane, A.B., Ladet, P., 2006: Operating room scheduling. International Journal of Productions Economics, 99, 52–62.
34. Murata,T., Ishibuchi, H., Tanaka,H., 1996: Multi-objective genetic algorithm and its application to flow shop scheduling, Comput. Ind. Eng. 30, 957–968.
35. Stacy, L.Janak., Xiaoxia, Lin., Christodoulos, A. Floudas, 2007: A new robust optimization approach for scheduling under uncertainty: II. Uncertainty with known probability distribution. Computers & Chemical Engineering, 31 (3), 171–195.
36. Chaari, T., Chaabane, S., Aissani, N., Trentesaux, D., 2014: Scheduling under uncertainty: Survey and research directions.3rd International Conference on Advanced Logistics and transport, IEEE. https://doi.org/10.1109/ICAdLT.2014.6866316.
37. Chaari, T., Chaabane, S., Loukil, T., Trentesaux, D., 2011: A genetic algorithm for robust hybrid flow shop scheduling, International Journal of Computer Integrated Manufacturing, Vol 24,N 9, pp 821–833.
38. Kouvelis, P., Yu,G., 1997: Robust discrete optimisation and its applications. Kluwer Academic Publisher, London.
39. Sevaux, M., Sorensen, K., 2004: A genetic algorithm for robust schedules in a just-in-time environment with ready times and due dates. 4OR - Quarterly journal of the Belgian, French and Italian Operations Research Societies, 2(2), 129–147.

40. Rajkumar, R., Shahabudeen, P., 2009: An improved genetic algorithm for the flowshop scheduling problem. International Journal of Production Research, 47 (1), 233–249.
41. Xiang, W., Yin, J., Lim, G., 2015: An ant colony optimization approach for solving an operating room surgery scheduling problem. Computer Industrial Engineering 85:335–345.
42. Santos, D.L., Hunsucker, J.L. , Deal, D.E., 1995: Global lower bounds for flow shops with multiple processors. European Journal of Operational Research, 80, 112–120.

Chapter 4
Cost Function Networks to Solve Large Computational Protein Design Problems

David Allouche, Sophie Barbe, Simon de Givry, George Katsirelos,
Yahia Lebbah, Samir Loudni, Abdelkader Ouali, Thomas Schiex,
David Simoncini, and Matthias Zytnicki

Abstract Proteins are chains of simple molecules called amino acids. The sequence of amino acids in the chain defines the three-dimensional shape of the protein and ultimately its biochemical function. Over millions of years, living organisms have evolved a large catalog of proteins. By exploring the space of possible amino acid sequences, protein engineering aims at similarly designing tailored proteins with specific desirable properties such as therapeutic properties in biomedical engineering for healthcare purposes. In computational protein design (CPD), the challenge of identifying a protein that performs a given task is defined as the combinatorial optimization of a complex energy function over amino acid sequences. First, we introduce the CPD problem and some of the main approaches that have been used by structural biologists to solve it. The CPD problem can be formulated as

D. Allouche · S. de Givry · T. Schiex (✉) · M. Zytnicki
Université Fédérale de Toulouse, ANITI, INRAE, Toulouse, France
e-mail: david.allouche@inrae.fr; simon.de.givry@inrae.fr; thomas.schiex@inrae.fr;
matthias.zytnicki@inrae.fr

S. Barbe
LISBP, INSA, UMR INRA 792/CNRS 5504, Toulouse, France
e-mail: sophie.barbe@insa-toulouse.fr

G. Katsirelos
UMR MIA-Paris, INRA, AgroParisTech, University of Paris-Saclay, Paris, France

Y. Lebbah
LITIO, University of Oran 1, Oran, Algeria

S. Loudni
GREYC, UMR 6072-CNRS, University of Caen Normandy, Caen, France
e-mail: samir.loudni@unicaen.fr

A. Ouali
ORPAILLEUR Team, LORIA, INRIA - Nancy, Vandœuvre-lès-Nancy, France
e-mail: abdelkader.ouali@inria.fr

D. Simoncini
IRIT, UMR 5505-CNRS, University of Toulouse, Toulouse, France
e-mail: david.simoncini@ut-capitole.fr

© Springer Nature Switzerland AG 2021   81
M. Masmoudi et al. (eds.), *Operations Research and Simulation
in Healthcare*, https://doi.org/10.1007/978-3-030-45223-0_4

a cost function network (CFN). We present some of the most efficient techniques in CFN. Overall, the CFN approach shows the best efficiency on these problems, improving by several orders of magnitude against the previous exact CPD-dedicated approaches and also against integer programming approaches.

4.1 Introduction

A protein is a sequence of basic building blocks called *amino acids*. There are 20 natural amino acids. Proteins are involved in nearly all structural, catalytic, sensory, and regulatory functions of living systems [21]. Performing these functions generally requires that proteins are assembled into well-defined three-dimensional structures specified by their amino acid sequence. Over millions of years, natural evolutionary processes have shaped and created proteins with novel structures and functions by means of sequence variations, including mutations, recombinations, and duplications. Protein engineering techniques coupled with high-throughput automated procedures make it possible to mimic the evolutionary process on a greatly accelerated time-scale and thus increase the odds to identify the proteins of interest for technological uses [62]. This holds great interest for medicine, synthetic biology, nanotechnologies, and biotechnologies [33, 56, 65]. In particular, protein engineering has become a key technology to generate tailored enzymes able to perform novel specific transformations under specific conditions. Such biochemical transformations enable to access a large repertoire of small molecules for various applications such as biofuels, chemical feedstocks, and therapeutics [8, 39]. The development of enzymes with required substrate selectivity, specificity, and stability can also be profitable to overcome some of the difficulties encountered in synthetic chemistry. In this field, the *in vitro* use of artificial enzymes in combination with organic chemistry has led to innovative and efficient routes for the production of high-value molecules while meeting the increasing demand for ecofriendly processes [10, 50, 74]. Nowadays, protein engineering is also being explored to create non-natural enzymes that can be combined *in vivo* with existing biosynthetic pathways or be used to create entirely new synthetic metabolic pathways not found in nature to access novel (bio)chemical products [25]. These latest approaches are central to the development of synthetic biology. One significant example in this field is the full-scale production of the antimalarial drug (artemisinin) from the engineered bacteria *Escherichia coli* [54].

With a choice among 20 naturally occurring amino acids at every position, the size of the combinatorial sequence space is out of reach for current experimental methods, even for short sequences. Computational protein design (CPD) methods, therefore, try to intelligently guide the protein design process by producing a *collection* of proteins that is rich in functional proteins, but small enough to be experimentally evaluated. The challenge of choosing a sequence of amino acids to perform a given task is formulated as an optimization problem, solvable computationally. It is often described as the inverse problem of protein folding [61]:

the three-dimensional structure is known and we have to find amino acid sequences that fold into it. It can also be considered as a highly combinatorial variant of side-chain positioning [70] because of possible amino acid mutations.

Various computational methods have been proposed over the years to solve this problem and several success stories have demonstrated the outstanding potential of CPD methods to engineer proteins with improved or novel properties. CPD has been successfully applied to increase protein thermostability and solubility, to alter specificity toward some other molecules, and to design various binding sites and construct *de novo* enzymes (see, for example, [40]).

Despite these significant advances, CPD methods must still mature in order to better guide and accelerate the construction of tailored proteins. In particular, more efficient computational optimization techniques are needed to explore the vast combinatorial space and to facilitate the incorporation of more realistic, flexible protein models. These methods need to be capable of not only identifying the optimal model, but also of enumerating solutions close to the optimum [73].

We begin by defining the CPD problem with a rigid backbone and then introduce the approach commonly used in structural biology to exactly solve CPD. This approach relies on dead-end elimination (DEE), a specific form of dominance analysis that was introduced in [19], and later strengthened in [31]. If this polynomial-time analysis does not solve the problem, an A^* algorithm is used to identify an optimal protein design.

We observe that the rigid backbone CPD problem can be naturally expressed as a cost function network (CFN) and solved as a weighted constraint satisfaction problem. In this context, DEE is similar to neighborhood substitutability [23, 29, 49].

To evaluate the efficiency of the CFN approach, we model the CPD problem using different combinatorial optimization formalisms. We compare the performance of the 0/1 linear programming solver cplex, and the CFN solver toulbar2, against that of a well-established CPD approach implementing DEE/A^*, on various realistic protein design problems. We observe drastic differences in the difficulty that these instances represent different solvers, despite often closely related models and solving techniques.

In Sect. 4.2, we describe the CPD problem and its previously- known solving methods. In Sect. 4.3, we show how to translate the CPD problem into a CFN in order to use CFN solving methods. In Sect. 4.4, we give an integer programming formulation of the CPD problem. In Sect. 4.5, we present our CPD benchmark instances. We report computational results in Sect. 4.6 and conclude.

4.2 The Computational Protein Design Approach

A protein is a sequence of organic compounds called amino acids. Each of the 20 amino acids consists of a common *peptidic core* and a *side chain* with varying chemical properties (see Fig. 4.1). In a protein, amino acid cores are linked together

Fig. 4.1 A representation of how amino acids, carrying specific side chains *R* and *R′*, can link together through their core to form a chain (modified from Wikipedia). One molecule of water is generated in the process

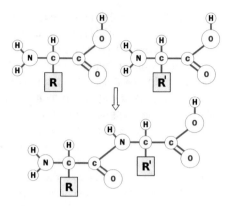

in sequence to form the *backbone* of the protein. In water, most proteins of interest *fold* into a 3D shape that is determined by the sequence of amino acids. Depending upon the amino acid considered, the side chain of each amino acid can be rotated along up to 4 dihedral angles relative to the backbone. Anfinsen's postulated [4] that the 3D structure of a protein is entirely defined by its sequence. This structure, defined by the backbone's structure and all side-chain rotations, is called the *conformation* of the protein and determines its chemical reactivity and biological function.

Computational protein design is faced with several challenges. The first lies in the exponential size of the conformational and protein sequence space that has to be explored, which rapidly grows out of reach of computational approaches. Another obstacle to overcome is the accurate structure prediction for a given sequence [32, 41]. Therefore, the design problem is usually approached as an inverse folding problem [61], in order to reduce the problem to the identification of an amino acid sequence that can fold into a target 3D-scaffold that matches the design objective [6]. In structural biology, the stability of a conformation can be directly evaluated through the energy of the conformation, a stable fold being of minimum energy [4].

In CPD, two approximations are common. First, it is assumed that the resulting designed protein retains the overall fold of the chosen scaffold: the protein *backbone* is considered fixed. At specific positions chosen automatically or by the molecular modeler, the amino acid used can be modified, thus changing the *side chain* as shown in Fig. 4.2. Second, the domain of conformations available to each amino acid side chain is continuous. This continuous domain is approximated using a set of discrete conformations defined by the value of their inner dihedral angles. These conformations, or *rotamers* [38], are derived from the most frequent conformations in the experimental repository of known protein structures, the PDB (Protein Data Bank, www.pdb.org). Different discretizations have been used in constraint-based approaches for protein structure prediction [7]. More recently, continuous optimization of dihedrals has been addressed in some limited settings [24, 34].

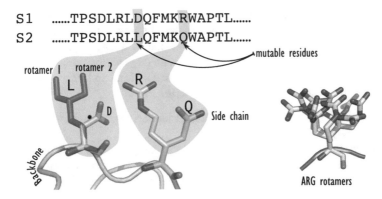

S1 ......TPSDLRLDQFMKRWAPTL......
S2 ......TPSDLRLLQFMKQWAPTL......

Fig. 4.2 A local view of the combinatorial sequence exploration considering a common backbone. Changes can be caused by amino acid identity substitutions (*for example, D/L or R/Q*) or by amino acid side-chain reorientations (rotamers) for a given amino acid. A sparse rotamer library for one amino acid is shown on the right (ARG=Arginine)

The CPD is then formulated as the problem of identifying a conformation of minimum energy via the mutation of a specific subset of amino acid residues, i.e. by affecting their identity and their 3D orientations (rotamers). The conformation that minimizes the energy is called the *GMEC* (Global Minimum Energy Conformation).

In order to solve this problem, we need a computationally tractable energetic model to evaluate the energy of any combination of rotamers. We also require computational optimization techniques that can efficiently explore the sequence-conformation space to find the sequence-conformation model of global minimum energy.

Energy Functions
Various energy functions have been defined to make energy computation manageable [5]. These energy functions include non-bonded terms such as van der Waals and electrostatic terms, often in conjunction with empirical contributions describing hydrogen bonds. The surrounding solvent effect is generally treated implicitly as a continuum. Statistical terms may be added in order to approximate the effect of mutations on the unfolded state or the contribution of conformational entropy. Finally, collisions between atoms (steric clashes) are also taken into account. We have used two state-of-the-art energy functions, AMBER9 [9] and Talaris2014 [58], respectively, implemented in the CPD-dedicated tools osprey 2.0 [26] and Rosetta Molecular Modeling suite 3 [48].

These energy functions are formulated in such a way that the terms are locally decomposable. Then, the energy of a given protein conformation, defined by a choice of one specific amino acid with an associated conformation (rotamer) for each residue, can be written as:

$$E = E_\varnothing + \sum_i E(i_r) + \sum_i \sum_{j>i} E(i_r, j_s) \tag{4.1}$$

where E is the potential energy of the protein, $E_\varnothing$ is a constant energy contribution capturing interactions between fixed parts of the model, $E(i_r)$ is the energy contribution of rotamer r at position i capturing internal interactions (and a reference energy for the associated amino acid) or interactions with fixed regions, and $E(i_r, j_s)$ is the pairwise interaction energy between rotamer r at position i and rotamer s at position j [19]. This decomposition brings two properties:

- Each term in the energy can be computed for each amino acid/rotamer (or pair for $E(i_r, j_s)$) independently.
- These energy terms, in $kcal/mol$, can be precomputed and cached, allowing to quickly compute the energy of a design once a specific rotamer (an amino acid-conformation pairing) has been chosen at each non-rigid position.

The rigid backbone discrete rotamer CPD problem is, therefore, defined by a fixed backbone with a corresponding set of positions (residues), a rotamer library, and a set of energy functions. Each position i of the backbone is associated with a subset D_i of all (amino acid, rotamer) pairs in the library. The problem is to identify at each position i a pair from D_i such that the overall energy E is minimized. In practice, based on expert knowledge or on specific design protocols, each position can be fixed (D_i is a singleton), flexible (all pairs in D_i have the same amino acid), or mutable (the general situation).

4.2.1   Exact CPD Methods

The protein design problem as defined above, with a rigid backbone, a discrete set of rotamers, and pairwise energy functions has been proven to be NP-hard [64]. Hence, a variety of meta-heuristics have been applied to it, including Monte Carlo simulated annealing [43], genetic algorithms [12, 66], variable neighborhood search [11], and other algorithms [20]. The main weakness of these approaches is that they may remain stuck in local minima and miss the GMEC without notice.

However, there are several important motivations for solving the CPD problem exactly. First, because they know when an optimum is reached, exact methods may stop before meta-heuristics. Voigt et al. [75] reported that the accuracy of meta-heuristics also degrades as problem size increases. More importantly, the use of exact search algorithms is important in the usual experimental design cycle that goes through modeling, solving, protein synthesis, and experimental evaluation: when unexpected experimental results are obtained, the only possible culprit lies in the CPD model and not in the algorithm.

Usual exact methods for CPD mainly rely on the dead-end elimination (DEE) theorem [18, 19] and the A^* algorithm [28, 47]. DEE is used as a preprocessing technique and removes rotamers that are locally dominated by other rotamers, until a fixpoint is reached. The rotamer r at position i (denoted by i_r) is removed if there exists another rotamer u at the same position such that [19]:

$$E(i_r) + \sum_{j \neq i} \min_s E(i_r, j_s) \geq E(i_u) + \sum_{j \neq i} \max_s E(i_u, j_s) \tag{4.2}$$

This condition guarantees that for any conformation with this r, we get a conformation with lower energy if we substitute u for r. Then, r can be removed from the list of possible rotamers at position i. This local dominance criterion was later improved by Goldstein [31] by directly comparing the energies of each rotamer in the same conformation:

$$E(i_r) - E(i_u) + \sum_{j \neq i} \min_s [E(i_r, j_s) - E(i_u, j_s)] \geq 0 \tag{4.3}$$

where the best and worst cases are replaced by the worst difference in energy. It is easy to see that this condition is always weaker than the previous one, and therefore, applicable to more cases. These two properties define polynomial-time algorithms that prune dominated values.

Since its introduction in 1992 by Desmet, DEE has become the fundamental tool of exact CPD, and various extensions have been proposed [27, 51, 63]. All these DEE criteria preserve the optimum but may remove suboptimal solutions. However CPD is NP-hard, and DEE cannot solve all CPD instances. Therefore, DEE preprocessing is usually followed by an A^* search. After DEE pruning, the A^* algorithm allows expanding a sequence-conformation tree, so that sequence-conformations are extracted and sorted on the basis of their energy values. The admissible heuristic (lower bound) used by A^* is described in [28].

When the DEE algorithm does not significantly reduce the search space, the A^* search tree can be too slow or memory demanding and the problem cannot be solved. Therefore, to circumvent these limitations and increase the ability of CPD to tackle problems with larger sequence-conformation spaces, novel alternative methods are needed. We now describe alternative state-of-the-art methods for solving the GMEC problem that offer attractive alternatives to DEE/A^*.

4.3 From CPD to CFN

CPD instances can be directly represented as cost function networks.

Definition 4.1 ([13]) A cost function network (CFN) is a pair (X, W), where $X = \{1, \ldots, n\}$ is a set of n variables and W is a set of cost functions. Each variable $i \in X$ has a finite domain D_i of values that can be assigned to it. A value $r \in D_i$ is denoted i_r. For a set of variables $S \subseteq X$, D_S denotes the Cartesian product of the domains of the variables in S. For a given tuple of values t, $t[S]$ denotes the projection of t over S. A cost function $w_S \in W$, with scope $S \subseteq X$, is a function $w_S : D_S \mapsto [0, k]$, where k is a maximum integer cost used for forbidden assignments.

We assume, without loss of generality, that every CFN includes at least one unary cost function w_i per variable $i \in X$ and a nullary cost function $w_\emptyset$. All costs being non-negative, the value of this constant function, $w_\emptyset$, provides a lower bound on the cost of any assignment.

The weighted constraint satisfaction problem (WCSP) is to find a complete assignment t minimizing the combined cost function $\bigoplus_{w_S \in W} w_S(t[S])$, where $a \oplus b = \min(k, a + b)$ is the k-bounded addition. This optimization problem has an associated NP-complete decision problem. Notice that if $k = 1$, then the WCSP is nothing but the famous constraint satisfaction problem or CSP (not the Max-CSP).

Modeling the CPD problem as a CFN is straightforward. The set of variables X has one variable i per residue i. The domain of each variable is the set of *(amino acid,conformation)* pairs in the rotamer library used. The global energy function can be represented by 0-ary, unary, and binary cost functions, capturing the constant energy term $w_\emptyset = E_\emptyset$, the unary energy terms $w_i(r) = E(i_r)$, and the binary energy terms $w_{ij}(r, s) = E(i_r, j_s)$, respectively. In the rest of the paper, for simplicity and consistency, we use notations $E_\emptyset$, $E(\cdot)$, and $E(\cdot, \cdot)$ to denote cost functions and restrict ourselves to binary CFN (extensions to higher orders are well-known).

Notice that there is one discrepancy between the original formulation and the CFN model: energies are represented as arbitrary floating-point numbers, while CFN uses positive costs. This can simply be fixed by first subtracting the minimum energy from all energy factors. These positive costs can then be multiplied by a large integer constant M and rounded to the nearest integer if integer costs are required.

4.3.1   Local Consistency in CFN

The usual exact approach to solve a CFN is to use a depth-first branch-and-bound algorithm (DFBB). A family of efficient and incrementally computed lower bounds is defined by local consistency properties.

Node consistency [44] (NC) requires that the domain of every variable i contains a value r that has a zero unary cost ($E(i_r) = 0$). This value is called the unary support for i. Furthermore, in the scope of the variable i, all values should have a cost below k ($\forall r \in D_i, E_\emptyset + E(i_r) < k$).

Soft arc consistency (AC*) [44, 68] requires NC and also that every value r of every variable i has a support on every cost function $E(i_r, j_s)$ involving i. A support of i_r is a value $j_s \in D_j$ such that $E(i_r, j_s) = 0$.

Stronger local consistencies such as existential directional arc consistency (EDAC) [45] and virtual arc consistency (VAC) [15] have also been introduced. See [13] for a review of existing local consistencies.

As in classical CSP, enforcing a local consistency property on a problem P involves transforming $P = (X, W)$ into a problem $P' = (X, W')$ that is equivalent to P (all complete assignments keep the same cost) and that satisfies the considered local consistency property. Enforcing a local consistency may increase $E_\emptyset$ and thus

Table 4.1 Time and space complexities of enforcing local consistency properties on a binary CFN (X, W), with $n = |X|$ variables, maximum domain size $d = \max_{i \in X} |D_i|$, $e = |W|$ cost functions with maximum forbidden cost k, and minimum non-zero rational cost $\epsilon \in]0, 1]$. Also, we report polynomial classes which are solved by these local consistencies

Local consistency	Time complexity	Space complexity	Polynomial classes
NC	$O(nd)$	$O(nd)$	–
AC*	$O(n^2 d^2 + ed^3)$	$O(ed)$	–
EDAC	$O(ed^2 \max(nd, k))$	$O(ed)$	*Tree-structures*
VAC$_\epsilon$	$O(ed^2 k / \epsilon)$	$O(ed)$	Tree-struct., submodular func.

improve the lower bound on the optimal cost. This bound is used to prune the search tree during DFBB.

Local consistency is enforced using *equivalence preserving transformations* (EPTs) that move costs between different cost functions [13–17, 44–46, 67, 68]. For example, a variable i violating the NC property because all its values i_r have a non-zero $E(i_r)$ cost can be made NC by subtracting the minimum cost from all $E(i_r)$ and adding this cost to $E_\varnothing$. The resulting network is equivalent to the original network, but it has an increased lower bound $E_\varnothing$. Far more complex sequences or sets of EPTs can be required to enforce other local consistencies [13].

Local consistency can be enforced in polynomial time and space as summarized in Table 4.1. During search, incrementality is preserved along any branch of the search tree, avoiding to repeatedly apply the same EPTs at every search node.

Unfortunately, finding the optimal *sequence* of EPTs with *integer* costs which maximizes $E_\varnothing$ is NP-hard [17]. It is polynomial for a *set* of EPTs with *rational* costs [16] and corresponds to optimization over the *local polytope* of probabilistic graphical models [13]. In practice, during search, applying suboptimal sequences of EPTs by enforcing EDAC or VAC is the most efficient approach for solving CFNs [37].

4.3.2 Maintaining Dead-End Elimination

Dead-end elimination is the key algorithmic tool of exact CPD solvers. From an AI perspective, in the context of CSP (if $k = 1$), the DEE equation (4.3) is equivalent to neighborhood substitutability [23]. In the context of CFN, the authors of [49] introduced partial soft neighborhood substitutability with a definition that is equivalent to Eq. (4.3) for pairwise decomposed energies.

DEE can be enforced in time $O(n^2 d^3)$ and it is orthogonal to local consistencies, except for VAC where value removals done by DEE cannot break the VAC property [49]. In practice, DEE and AC* (or EDAC) will be enforced until both properties are verified, with time complexity in $O(n^3 d^4)$. In order to reduce this time complexity, during search, we do not test every pair of values in every domain D_i, but only one pair (i_r, i_u) such that $E(i_u)$ is minimum and $E(i_r)$ is maximum. Thus, enforcing this restricted DEE[1] can be done in $O(n^2 d)$ and it iterates at most $n \times d$ times with AC* [29].

4.3.3 Exploiting Tree Decomposition in a Hybrid Best-First Branch-and-Bound Method

Hybrid best-first search (HBFS) [2] explores the search tree in a best-first manner as in A*. However, each selected open search node is expanded by a depth-first search with a limited number of backtracks. When the limit is reached, all the remaining unexplored search nodes are inserted in the open node list. The limit is dynamically tuned in order to reduce the overhead of reconstructing the CFN corresponding to the selected open node.

The HBFS method was further extended in [2] to exploit a tree decomposition, resulting in the BTD-HBFS method.

Definition 4.2 A tree decomposition of a connected CFN (X, W) is a pair (C_T, T), where $T = (I, A)$ is a tree with nodes set I and edges set A and $C_T = \{C_i \mid i \in I\}$ is a family of subsets of X, called *clusters*, such that: (i) $\cup_{i \in I} C_i = X$, (ii) $\forall w_S \in W$, $\exists C_i \in C_T$ s.t. $S \subseteq C_i$, (iii) $\forall i, j, k \in I$, if j is on the path from i to k in T, then $C_i \cap C_k \subseteq C_j$.

Definition 4.3 A graph of clusters for a tree decomposition (C_T, T) is an undirected graph $G = (C_T, E)$ that has a vertex for each cluster $C_i \in C_T$, and there is an edge $(C_i, C_j) \in E$ when $C_i \cap C_j \neq \emptyset$.

The width of a tree decomposition corresponds to the size of the largest cluster minus one. As finding an optimal tree decomposition with minimum width, called *treewidth*, is NP-hard, we use fast approximate algorithms like the *min-fill* heuristic.

BTD-HBFS (*backtracking with tree decomposition and HBFS*) uses a restricted variable ordering heuristic, which selects variables from a root (largest) cluster first and then continues by assigning variables in the child clusters in a depth-first manner. Each child cluster has one open node list with associated lower and upper bounds per visited assignment of its variables intersecting with its parent cluster. By doing so, it exploits the lower bounds reported by HBFS in individual clusters to improve the anytime behavior and global pruning of BTD. Given a CFN (X, W) with treewidth t, BTD computes the optimum in time $O(knd^{t+1})$ and space $O(knd^{2t})$ [2, 30, 71].

4.3.4 A Parallel Variable Neighborhood Search Method Guided by Tree Decomposition

UDGVNS (for unified decomposition guided variable neighborhood search) [59, 60] is a CFN solving method unifying two complete and incomplete search methods: iterative limited discrepancy search (LDS) and variable neighborhood search (VNS). LDS [35] is a heuristic method that explores the depth-first search tree in a non-systematic way by making a limited number of *wrong* decisions w.r.t. its value ordering heuristic. We assume a binary search tree where at each search node either the selected variable is assigned to its chosen preferred value (left branch) or the value is removed from the domain (right branch). Each value removal corresponds to a wrong decision made by the search, it is called a *discrepancy*. The number of discrepancies is limited by a parameter. Iterative LDS increases this parameter (by a multiplication factor of 2) at each iteration until a complete search is done.

VNS/LDS [52, 55] is a meta-heuristic that uses a finite set of preselected neighborhood structures $N_s, s = 4, 5, \ldots, n$ to escape from local minima by systematically changing the neighborhood structure if the current one does not improve the incumbent solution. The initial solution is found by a depth-first search on the whole problem. Then, s variables are randomly chosen and the current solution is partially destroyed by unassigning the selected variables and an exploration of its (large) neighborhood is performed by LDS with a fixed discrepancy. As soon as a better solution is found, then s is reset to its minimal value (4 in our experiments). DGVNS [22] uses another neighborhood structure $N_{s,c}$, where s is the neighborhood size and C_c is the cluster where the variables will be selected from. If $s > |C_c|$, we complete the set of candidate variables to be unassigned by adding clusters adjacent to C_c in the graph of clusters provided by a tree decomposition. The neighborhood change in DGVNS is performed in the same way as in VNS/LDS. However, DGVNS considers successively all the clusters. This ensures a better diversification by covering a large number of different regions. UDGVNS [59, 60] iterates DGVNS with an increasing discrepancy (when $s = n$ and no better solution was found) as in iterative LDS, until a complete search is done, therefore, tuning its compromise between optimality proof and anytime behavior.

The parallel version of UDGVNS relies on a master/worker model. The master process controls the communication over all the processes and holds the centralized information, while the asynchronous worker processes explore the parts of the search space assigned by the master. The cooperation in parallel UDGVNS is achieved by sharing asynchronously a single global best solution among the worker processes.

4.4   Integer Linear Programming for the CPD

The rigid backbone CPD problem has a simple formulation and can be easily written in a variety of combinatorial optimization frameworks. In our previous work [1], we compared the CFN approach to solvers coming from different fields, including probabilistic graphical model, 0/1 linear programming, 0/1 quadratic programming, and 0/1 quadratic optimization. Here, we present only the 01LP model. Other approaches were shown to be far less efficient than CFN and 01LP [1].

A 0/1 linear programming (01LP) problem is defined by a linear criterion to optimize over a set of Boolean variables under a conjunction of linear inequalities and equalities.

For every assignment i_r of every variable i, there is a Boolean variable d_{ir} that is equal to 1 iff rotamer r is chosen at position i. Additional constraints (4.4) enforce that exactly one value is selected for each variable. For every pair of values of different variables (i_r, j_s) involved in a binary energy term, there is a Boolean variable p_{irjs} that is equal to 1 iff the pair (i_r, j_s) is used. Constraints (4.5) enforce that a pair is used iff the corresponding values are used. Then, finding a GMEC reduces to the following 01LP:

$$\min E_\varnothing + \sum_{\substack{i,r \\ E(i_r)\neq k}} E(i_r).d_{ir} + \sum_{\substack{i,r,j,s \\ j>i, E(i_r,j_s)\neq k}} E(i_r, j_s).p_{irjs}$$

$$\text{s.t.} \quad \sum_r d_{ir} = 1 \quad (\forall i) \tag{4.4}$$

$$\sum_s p_{irjs} = d_{ir} \quad (\forall i, r, j) \tag{4.5}$$

$$d_{ir} = 0 \quad (\forall i, r \quad \text{s.t.} \quad E(i_r) = k) \tag{4.6}$$

$$p_{irjs} = 0 \quad (\forall i, r, j, s \quad \text{s.t.} \quad E(i_r, j_s) = k) \tag{4.7}$$

$$d_{ir} \in \{0, 1\} \quad (\forall i, r) \tag{4.8}$$

$$p_{irjs} \in \{0, 1\} \quad (\forall i, r, j, s) \tag{4.9}$$

It is known that the continuous LP relaxation of this model is the dual of the LP problem defined by optimal soft arc consistency (OSAC) [13, 16] when the upper bound k used in CFN is infinite. OSAC is known to be stronger than any other soft arc consistency level, including EDAC and VAC. However, as soon as the upper bound k used for pruning in CFN decreases to a finite value, soft local consistencies may prune values and EDAC becomes incomparable with the dual of these relaxed LPs.

This model is also the 01LP model IP1 proposed in [42] for side-chain positioning. It has a quadratic number of Boolean variables. Constraints (4.6) and (4.7) explicitly forbid values and pairs with cost k (steric clashes).

This model can be simplified by relaxing the integrality constraint on the p_{irjs}: indeed, if all d_{ir} are set to 0 or 1 by constraint (4.8), the constraints (4.4) and (4.5) enforce that the p_{irjs} are set to 0 or 1. In the rest of the paper, we relax constraint (4.9).

4.5 Computational Protein Design Instances

In our initial experiments with CPD in [3], we built 12 designs using the CPD-dedicated tool osprey 1.0. A new version of osprey being available since, we used this new 2.0 version [26] for all computations. Among different changes, this new version uses a modified energy field that includes a new definition of the "reference energy" and a different rotamer library. We, therefore, rebuilt the 12 instances from [3] and additionally created 35 extra instances from existing published designs, as described in [72]. We must insist on the fact that the 12 rebuilt instances do not define the same energy landscape or search space as the initial [3]'s instances (due to changes in rotamers set).

These designs include protein structures derived from the PDB that were chosen for the high resolution of their 3D-structures, their use in the literature, and their distribution of sizes and types. Diverse sizes of sequence-conformation combinatorial spaces are represented, varying by the number of mutable residues, the number of alternative amino acid types at each position, and the number of conformations for each amino acid. The *Penultimate* rotamer library was used [53].

Preparation of CPD Instances
Missing heavy atoms in crystal structures and hydrogen atoms were added with the *tleap* module of the AMBER9 software package [9]. Each molecular system was then minimized in implicit solvent (Generalized Born model [36]) using the *Sander* program and the all-atom *ff99* force field of AMBER9. All $E_\varnothing$, $E(i_r)$, and $E(i_r, j_s)$ energies of rotamers (see Eq. (4.1)) were precomputed using osprey 2.0. The energy function consisted of the Amber electrostatic, van der Waals, and the solvent terms. Rotamers and rotamer pairs leading to steric clashes between molecules are associated with huge energies (10^{38}) representing forbidden combinations. For n residues to optimize with d possible (amino acid,conformation) pairs, there are n unary and up to $\frac{n.(n-1)}{2}$ binary cost functions that can be computed independently. The resulting instances have between $n = 11$ and $n = 120$ residues and between $d = 48$ and $d = 198$ rotamers (see Table 4.2).

Translation to WCSP and 01LP Formats
The native CPD problems were translated to the WCSP format before any preprocessing. To convert the floating-point energies of a given instance to non-negative integer costs, we subtracted the minimum energy to all energies and then multiplied energies by an integer constant M and rounded to the nearest integer. The initial upper bound k is set to the sum, over all cost functions, of the maximum energies (excluding forbidden steric clashes). High energies corresponding to steric clashes

are represented as costs equal to the upper bound k (the forbidden cost). The resulting model was used as the basis for all other solvers (except `osprey`). To keep a cost magnitude compatible with all the compared solvers, we used $M = 10^2$. Experiments with a finer discretization ($M = 10^8$) were used in previous experiments [72] with no significant difference in computing efforts.

Note that the approximate treewidth produced by the min-fill heuristic of our WCSP instances, after preprocessing by `toulbar2`, is between 4 and 113 (see Table 4.2).

All the Python and C translating scripts used, including translating from WCSP to 01LP format, are available together with the 47 CPD instances in native and WCSP formats (1.7GB) at http://genoweb.toulouse.inra.fr/~tschiex/CPD-AIJ.

4.6   Experimental Results

For computing the GMEC, all computations were performed on a dual-socket 12-core-per-socket Intel Xeon E5-2680 v3 at 2.5 GHz, 256 GB of RAM (except `osprey` on an AMD Operon 6176 at 2.3 GHz, 128 GB of RAM), and a 9,000-second time-out. We compared three solvers: DEE/A* `osprey` version 2.0 (www.cs.duke.edu/donaldlab/osprey.php), 01LP solver `cplex` version 12.8.0 (with parameters EPAGAP, EPGAP, and EPINT set to zero to avoid premature stop), and CFN solver `toulbar2` version 1.0.1 (www.github.com/toulbar2/toulbar2).

The procedure of `osprey` starts with extensive DEE preprocessing (*algOption* = 3, includes simple Goldstein, Magic bullet pairs, 1 and 2-split positions, Bounds and pairs pruning) followed by A* search. Only the GMEC conformation is generated by A* (*initEw*=0).

We ran `toulbar2` using VAC in preprocessing and HBFS with EDAC, DEE[1], and binary branching during search (options -d: -A). The additional exploitation of a min-fill tree decomposition inside branch-and-bound (see BTD-HBFS in Sect. 4.3.3) is denoted as `toulbar2` BTD (extra options -O=-3 -B=1). The variable neighborhood search guided by tree decomposition is denoted `toulbar2` UDGVNS (extra options -O=-3 -vns).

Table 4.2 and Fig. 4.3.left report the number of problems solved within a given time limit. `osprey` solved 25 instances, `cplex` solved 30 (resp. 33 with using 10 cores), `toulbar2` solved 40 (resp. 39 with UDGVNS). 7 instances remain unsolved by any approaches in less than 9,000 s. Within the three CFN approaches, `toulbar2` BTD is the fastest in terms of optimality proofs. Using 10 cores improves `cplex` results by solving 3 more instances and a mean speed-up factor of 2.2 on the same 30 solved instances. A different behavior was found for `toulbar2` UDGVNS where the mean speed-up factor was only 1.15 in favor of the parallel version. This is because the parallelization of `toulbar2` UDGVNS is to partially explore the neighborhood in parallel until a single process explores the whole problem sequentially using a complete depth-first branch-and-bound and ends the search.

Table 4.2 For each instance: protein (PDB id.), number of mutable residues, maximum domain size (maximum number of rotamers), approximate treewidth, and CPU-time for solving (with best solution found in parentheses), a bold number indicates that the solver found the best solution among all solvers) using cplex, cplex (10 cores), toulbar2, toulbar2 BTD, toulbar2 UDGVNS, toulbar2 UDGVNS (10 cores). A "-" indicates that the corresponding solver did not prove optimality within the 9,000-second time-out. A "!" indicates the solver stops with a SEGV signal

PDB id.	n	d	t	cplex	cplex-10	toulbar2	toulbar2 BTD	tb2 UDGVNS	UDGVNS	tb2 UDGVNS-10
2TRX	11	48	10	1.7 **(178440)**	2.1 **(178440)**	**0.1 (178440)**	**0.1 (178440)**	**0.1 (178440)**	**0.1 (178440)**	**0.1 (178440)**
1PGB	11	49	9	2.0 **(125306)**	3.8 **(125306)**	**0.1 (125306)**	**0.1 (125306)**	**0.1 (125306)**	**0.1 (125306)**	**0.1 (125306)**
1PGB	11	198	10	- (286299)	2781 **(286135)**	**2.5 (286135)**	**2.5 (286135)**	4.1 **(286135)**	4.0 **(286135)**	4.0 **(286135)**
1HZ5	12	49	10	5.1 **(150714)**	4.5 **(150714)**	**0.1 (150714)**	**0.1 (150714)**	**0.1 (150714)**	**0.1 (150714)**	**0.1 (150714)**
1HZ5	12	198	11	635.1 **(342476)**	264.0 **(342476)**	1.4 **(342476)**	2.1 **(342476)**	2.1 **(342476)**	1.5 **(342476)**	1.5 **(342476)**
1UBI	13	49	12	77.7 **(159522)**	34.3 **(159522)**	**0.3 (159522)**	**0.3 (159522)**	**0.3 (159522)**	**0.3 (159522)**	**0.3 (159522)**
1UBI	13	198	12	- (383687)	- (382996)	1023 **(380554)**	663.2 **(380554)**	20.4 **(380554)**	- **(380554)**	- **(380554)**
2DHC	14	198	13	- (1410934)	- (1411556)	8.2 **(1410850)**	6.1 **(1410850)**	20.4 **(1410850)**	17.2 **(1410850)**	17.2 **(1410850)**
1CM1	17	198	15	138.4 **(743645)**	116.6 **(743645)**	1.8 **(743645)**	**1.7 (743645)**	3.2 **(743645)**	**1.7 (743645)**	**1.7 (743645)**
2PCY	18	48	11	12.8 **(307667)**	14.5 **(307667)**	**0.2 (307667)**	**0.3 (307667)**	**0.2 (307667)**	**0.3 (307667)**	**0.3 (307667)**
1PIN	28	198	27	- (1999094)	- (1999094)	- (1994157)	- (1994179)	- (1994135)	- **(1994069)**	- **(1994069)**
1C9O	55	198	54	- (8024505)	! (8024505)	- (8014215)	- (8014405)	- (8013870)	- **(8013542)**	- **(8013542)**
1FYN	23	186	17	1235 **(1183427)**	902.3 **(1183427)**	1.7 **(1183427)**	2.6 **(1183427)**	2.3 **(1183427)**	1.6 **(1183427)**	1.6 **(1183427)**
1BK2	24	182	16	67.2 **(1133737)**	56.3 **(1133737)**	**0.3 (1133737)**	0.6 **(1133737)**	0.6 **(1133737)**	0.4 **(1133737)**	0.4 **(1133737)**
1MJC	28	182	4	1.8 **(1514364)**	2.4 **(1514364)**	**0.0 (1514364)**	**0.1 (1514364)**	**0.1 (1514364)**	**0.1 (1514364)**	**0.1 (1514364)**
1SHG	28	182	13	16.3 **(1513151)**	18.9 **(1513151)**	**0.1 (1513151)**	0.3 **(1513151)**	0.3 **(1513151)**	0.2 **(1513151)**	0.2 **(1513151)**
1PIN	28	194	20	7077 **(1570593)**	881.1 **(1570593)**	2.0 **(1570593)**	1.3 **(1570593)**	3.1 **(1570593)**	2.8 **(1570593)**	2.8 **(1570593)**
1CSK	30	49	13	8.1 **(1125798)**	9.0 **(1125798)**	**0.1 (1125798)**	**0.2 (1125798)**	**0.1 (1125798)**	**0.1 (1125798)**	**0.1 (1125798)**
1SHF	30	56	15	5.7 **(1101033)**	7.8 **(1101033)**	**0.1 (1101033)**	**0.2 (1101033)**	**0.2 (1101033)**	**0.1 (1101033)**	**0.1 (1101033)**
1CSP	30	182	10	964.2 **(2520706)**	120.4 **(2520706)**	**0.4 (2520706)**	**0.4 (2520706)**	0.7 **(2520706)**	0.7 **(2520706)**	0.7 **(2520706)**
1PGB	31	182	29	- (2442440)	- (2439846)	- **(2433714)**	- (2433843)	- **(2433714)**	- **(2433714)**	- **(2433714)**
1NXB	34	56	18	10.9 **(2971624)**	11.6 **(2971624)**	**0.1 (2971624)**	**0.2 (2971624)**	0.3 **(2971624)**	0.2 **(2971624)**	0.2 **(2971624)**

(continued)

Table 4.2 (continued)

PDB id.	n	d	t	cplex	cplex-10	toulbar2	toulbar2 BTD	tb2 UDGVNS	tb2 UDGVNS-10
1ENH	36	182	34	- (2637033)	- (2571387)	- (2571065)	- (2570942)	- (2570849)	- (2570849)
2DRI	37	186	24	- (2905652)	- (2905276)	135.5 (2905276)	19.2 (2905276)	762.0 (2905276)	1005 (2905276)
1FNA	38	48	19	63.5 (3750256)	85.5 (3750256)	0.4 (3750256)	0.5 (3750256)	0.5 (3750256)	0.4 (3750256)
1CTF	39	56	27	160.2 (1881332)	158.0 (1881332)	0.7 (1881332)	0.7 (1881332)	1.1 (1881332)	1.1 (1881332)
1TEN	39	66	11	19.1 (1959862)	17.6 (1959862)	0.1 (1959862)	0.3 (1959862)	0.2 (1959862)	0.2 (1959862)
1UBI	40	182	26	518.1 (3068938)	341.5 (3068938)	1.6 (3068938)	1.3 (3068938)	2.2 (3068938)	2.7 (3068938)
1CDL	40	186	33	- (N/A)	- (3594204)	392.6 (3590514)	233.3 (3590514)	1559 (3590514)	1366 (3590514)
1CM1	42	186	35	- (3973849)	6177 (3895415)	6.6 (3895415)	6.4 (3895415)	7.7 (3895415)	7.0 (3895415)
1C9O	43	182	26	885.9 (4959931)	387.4 (4959931)	1.1 (4959931)	1.5 (4959931)	1.4 (4959931)	1.2 (4959931)
1BRS	44	194	29	- (N/A)	- (4008249)	555.3 (4007610)	198.6 (4007610)	1371 (4007610)	1327 (4007610)
2PCY	46	56	20	33.4 (2935820)	32.7 (2935820)	0.2 (2935820)	0.5 (2935820)	0.5 (2935820)	0.3 (2935820)
1POH	46	182	18	17.6 (4033880)	25.9 (4033880)	0.2 (4033880)	0.2 (4033880)	0.3 (4033880)	0.3 (4033880)
1DKT	46	190	35	1634 (4192582)	1068 (4192582)	1.7 (4192582)	1.5 (4192582)	2.2 (4192582)	1.8 (4192582)
2CI2	51	183	41	- (N/A)	- (6391285)	- (6299287)	- (6299277)	- (6299194)	- (6299194)
1GVP	52	182	42	- (N/A)	- (5197032)	596.1 (5196719)	217.6 (5196719)	2469 (5196719)	1772 (5196719)
1RIS	56	182	40	- (6222964)	- (6171247)	129.7 (6171191)	21.6 (6171191)	77.0 (6171191)	84.2 (6171191)
1LZ1	59	57	33	581.6 (7022658)	292.7 (7022658)	1.5 (7022658)	1.5 (7022658)	1.8 (7022658)	2.3 (7022658)
2TRX	61	186	35	443.2 (7016169)	374.2 (7016169)	0.8 (7016169)	1.0 (7016169)	1.1 (7016169)	1.1 (7016169)
2RN2	69	66	38	210.2 (8909892)	257.4 (8909892)	1.1 (8909892)	1.6 (8909892)	1.4 (8909892)	2.0 (8909892)
3CHY	74	66	46	- (10461250)	5259 (10461151)	88.7 (10461151)	36.9 (10461151)	76.0 (10461151)	88.9 (10461151)
1L63	83	182	48	1602 (12891031)	1056 (12891031)	1.7 (12891031)	2.6 (12891031)	2.5 (12891031)	2.3 (12891031)
1HNG	85	182	29	1760 (13532638)	1672 (13532638)	2.0 (13532638)	2.6 (13532638)	3.2 (13532638)	3.6 (13532638)
1CSE	97	183	18	130.3 (18602292)	86.0 (18602292)	0.4 (18602292)	0.8 (18602292)	0.8 (18602292)	0.7 (18602292)
3HHR	115	186	85	- (N/A)	! (N/A)	- (89193514)	- (89191561)	- (89188633)	- (89188543)
1STN	120	190	113	- (37111502)	! (37111502)	- (37068565)	- (37070863)	- (37055381)	- (37054896)
				30 (30)	33 (34)	40 (41)	40 (40)	39 (43)	39 (47)

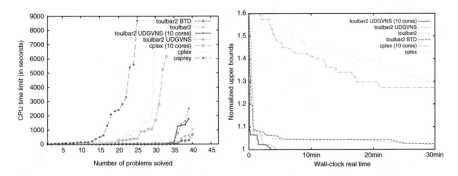

Fig. 4.3 (Left) A figure showing the number of problems that can be solved by each approach (X-axis) as a function of time allowed for solving each problem (Y-axis). (Right) A figure showing normalized upper bounds (Y-axis) on 47 instances as time passes (X-axis)

Figure 4.3.right compares the evolution of upper bounds as time passes for the different methods (except for osprey which returns one optimal solution at the end of its A^* exploration). Specifically, for each instance, we normalize all energies as follows: the best, potentially suboptimal solution found by any algorithm is 1, the worst solution is 2. This normalization is invariant to translation and scaling. Again, there is a clear separation between CFN and 01LP methods, in favor of CFN approaches. Among the sequential methods, toulbar2 UDGVNS has the best anytime profile. Its parallel version toulbar2 UDGVNS using 10 cores found the best upper bounds for all the 47 instances (see Table 4.2).

4.7   Conclusions

The simplest formal optimization problem underlying CPD looks for a global minimum energy conformation (GMEC) over a rigid backbone and altered side chains (identity and conformation). In computational biology, exact methods for solving the CPD problem combine dominance analysis (DEE) and an A^* search.

The CPD problem can also be directly formulated as a cost function network, with a (very) dense graph and relatively large domains. We have shown how DEE can be integrated with local consistency with reasonable time complexity. An alternative 01LP formulation leads to a much larger number of variables.

On a variety of real instances, we have shown that state-of-the-art optimization algorithms on integer programming or cost function network give important speedups compared to usual CPD algorithms combining dead-end elimination with $A*$. Among all the tested solvers, toulbar2 was the most efficient solver and its efficiency was further improved by exploiting a tree decomposition either for optimality proofs or for finding good anytime solutions on the most difficult instances. In practice, the performance of toulbar2 relies mostly on its fast lower bounds, while adding DEE has a marginal effect [1]. `toulbar2` has now been integrated inside the `osprey` CPD software and is being integrated inside `Rosetta Molecular Modeling suite` [48]. Indeed, another comparison has shown the Rosetta CPD-dedicated simulated annealing implementation was outperformed by the exact CFN `toulbar2` BTD approach [69] in terms of GMEC solution quality within a 100-h time limit. Exact optimization may also be useful in practice as we have recently shown using `toulbar2` for the design of a real self-assembling protein [57].

We also showed that these CPD problems define challenging benchmarks for AI and OR researchers. Among 47 instances, 7 remains open under 9,000 s, having from 28 to 120 mutable residues and at most 198 rotamers per position. Larger problems may exist as most of the known proteins in the PDB database include more than 200 residues.[1]

In practice, it must be stressed that just finding the GMEC is not a final answer to real CPD problems. CPD energies functions represent an approximation of the real physics of proteins and optimizing a target score based on them (such as stability, affinity, etc.) is not a guarantee of finding a successful design. Indeed, some designs may be so stable that they are unable to accomplish the intended biological function. The usual approach is, therefore, to design a large library of proteins whose sequences are extracted from all solutions within a small threshold of the energy of the GMEC. This problem is also efficiently solved by `toulbar2` [69, 72].

Acknowledgments This work has been partly funded by the "Agence nationale de la Recherche" (ANR-10-BLA-0214, ANR-12-MONU-0015-03, and ANR-16-CE40-0028).

References

1. Allouche, D., André, I., Barbe, S., Davies, J., de Givry, S., Katsirelos, G., O'Sullivan, B., Prestwich, S., Schiex, T., Traoré, S.: Computational protein design as an optimization problem. Artificial Intelligence **212**, 59–79 (2014)
2. Allouche, D., de Givry, S., Katsirelos, G., Schiex, T., Zytnicki, M.: Anytime Hybrid Best-First Search with Tree Decomposition for Weighted CSP. In: Proc. of CP-15, pp. 12–28. Cork, Ireland (2015)
3. Allouche, D., Traoré, S., André, I., de Givry, S., Katsirelos, G., Barbe, S., Schiex, T.: Computational protein design as a cost function network optimization problem. In: Principles and Practice of Constraint Programming, pp. 840–849. Springer (2012)
4. Anfinsen, C.: Principles that govern the folding of protein chains. Science **181**(4096), 223–253 (1973)
5. Boas, F.E., Harbury, P.B.: Potential energy functions for protein design. Current opinion in structural biology **17**(2), 199–204 (2007)
6. Bowie, J.U., Luthy, R., Eisenberg, D.: A method to identify protein sequences that fold into a known three-dimensional structure. Science **253**(5016), 164–170 (1991)

---

[1] www.rcsb.org/stats/distribution_residue-count.

7. Campeotto, F., Dal Palù, A., Dovier, A., Fioretto, F., Pontelli, E.: A constraint solver for flexible protein models. J. Artif. Int. Res. (JAIR) **48**(1), 953–1000 (2013)
8. Carothers, J.M., Goler, J.A., Keasling, J.D.: Chemical synthesis using synthetic biology. Current opinion in biotechnology **20**(4), 498–503 (2009)
9. Case, D., Darden, T., Cheatham III, T., Simmerling, C., Wang, J., Duke, R., Luo, R., Merz, K., Pearlman, D., Crowley, M., Walker, R., Zhang, W., Wang, B., Hayik, S., Roitberg, A., Seabra, G., Wong, K., Paesani, F., Wu, X., Brozell, S., Tsui, V., Gohlke, H., Yang, L., Tan, C., Mongan, J., Hornak, V., Cui, G., Beroza, P., Mathews, D., Schafmeister, C., Ross, W., Kollman, P.: Amber 9. Tech. rep., University of California, San Francisco (2006)
10. Champion, E., André, I., Moulis, C., Boutet, J., Descroix, K., Morel, S., Monsan, P., Mulard, L.A., Remaud-Siméon, M.: Design of α-transglucosidases of controlled specificity for programmed chemoenzymatic synthesis of antigenic oligosaccharides. Journal of the American Chemical Society **131**(21), 7379–7389 (2009)
11. Charpentier, A., Mignon, D., Barbe, S., Cortes, J., Schiex, T., Simonson, T., Allouche, D.: Variable neighborhood search with cost function networks to solve large computational protein design problems. Journal of Chemical Information and Modeling **59**(1), 127–136 (2019)
12. Chowdry, A.B., Reynolds, K.A., Hanes, M.S., Voorhies, M., Pokala, N., Handel, T.M.: An object-oriented library for computational protein design. J. Comput. Chem. **28**(14), 2378–2388 (2007)
13. Cooper, M., de Givry, S., Sanchez, M., Schiex, T., Zytnicki, M., Werner, T.: Soft arc consistency revisited. Artificial Intelligence **174**, 449–478 (2010)
14. Cooper, M.C.: High-order consistency in Valued Constraint Satisfaction. Constraints **10**, 283–305 (2005)
15. Cooper, M.C., de Givry, S., Sánchez, M., Schiex, T., Zytnicki, M.: Virtual arc consistency for weighted CSP. In: Proc. of AAAI'08, vol. 8, pp. 253–258. Chicago, IL (2008)
16. Cooper, M.C., de Givry, S., Schiex, T.: Optimal soft arc consistency. In: Proc. of IJCAI'2007, pp. 68–73. Hyderabad, India (2007)
17. Cooper, M.C., Schiex, T.: Arc consistency for soft constraints. Artificial Intelligence **154**(1-2), 199–227 (2004)
18. Dahiyat, B.I., Mayo, S.L.: Protein design automation. Protein science **5**(5), 895–903 (1996)
19. Desmet, J., De Maeyer, M., Hazes, B., Lasters, I.: The dead-end elimination theorem and its use in protein side-chain positioning. Nature **356**(6369), 539–42 (1992)
20. Desmet, J., Spriet, J., Lasters, I.: Fast and accurate side-chain topology and energy refinement (FASTER) as a new method for protein structure optimization. Proteins **48**(1), 31–43 (2002)
21. Fersht, A.: Structure and mechanism in protein science: a guide to enzyme catalysis and protein folding. WH. Freeman and Co., New York (1999)
22. Fontaine, M., Loudni, S., Boizumault, P.: Exploiting tree decomposition for guiding neighborhoods exploration for VNS. RAIRO OR **47**(2), 91–123 (2013)
23. Freuder, E.C.: Eliminating interchangeable values in constraint satisfaction problems. In: Proc. of AAAI'91, pp. 227–233. Anaheim, CA (1991)
24. Friesen, A.L., Domingos, P.: Recursive decomposition for nonconvex optimization. In: Proc. of IJCAI'15, pp. 253–259. Buenos Aires, Argentina (2015)
25. Fritz, B.R., Timmerman, L.E., Daringer, N.M., Leonard, J.N., Jewett, M.C.: Biology by design: from top to bottom and back. BioMed Research International **2010** (2010)
26. Gainza, P., Roberts, K.E., Georgiev, I., Lilien, R.H., Keedy, D.A., Chen, C.Y., Reza, F., Anderson, A.C., Richardson, D.C., Richardson, J.S., et al.: Osprey: Protein design with ensembles, flexibility, and provable algorithms. Methods Enzymol (2012)
27. Georgiev, I., Lilien, R.H., Donald, B.R.: Improved Pruning algorithms and Divide-and-Conquer strategies for Dead-End Elimination, with application to protein design. Bioinformatics **22**(14), e174–83 (2006)
28. Georgiev, I., Lilien, R.H., Donald, B.R.: The minimized dead-end elimination criterion and its application to protein redesign in a hybrid scoring and search algorithm for computing partition functions over molecular ensembles. Journal of computational chemistry **29**(10), 1527–42 (2008)

29. de Givry, S., Prestwich, S., O'Sullivan, B.: Dead-End Elimination for Weighted CSP. In: Proc. of CP-13, pp. 263–272. Uppsala, Sweden (2013)
30. de Givry, S., Schiex, T., Verfaillie, G.: Exploiting Tree Decomposition and Soft Local Consistency in Weighted CSP. In: Proc. of AAAI'06, pp. 22–27. Boston, MA (2006)
31. Goldstein, R.F.: Efficient rotamer elimination applied to protein side-chains and related spin glasses. Biophysical journal 66(5), 1335–40 (1994)
32. Gront, D., Kulp, D.W., Vernon, R.M., Strauss, C.E., Baker, D.: Generalized fragment picking in Rosetta: design, protocols and applications. PloS one 6(8), e23294 (2011)
33. Grunwald, I., Rischka, K., Kast, S.M., Scheibel, T., Bargel, H.: Mimicking biopolymers on a molecular scale: nano(bio)technology based on engineered proteins. Philosophical transactions. Series A, Mathematical, physical, and engineering sciences 367(1894), 1727–47 (2009)
34. Hallen, M.A., Keedy, D.A., Donald, B.R.: Dead-end elimination with perturbations (deeper): A provable protein design algorithm with continuous sidechain and backbone flexibility. Proteins: Structure, Function, and Bioinformatics 81(1), 18–39 (2013)
35. Harvey, W.D., Ginsberg, M.L.: Limited discrepancy search. In: Proc. of IJCAI'95. Montréal, Canada (1995)
36. Hawkins, G., Cramer, C., Truhlar, D.: Parametrized models of aqueous free energies of solvation based on pairwise descreening of solute atomic charges from a dielectric medium. The Journal of Physical Chemistry 100(51), 19824–19839 (1996)
37. Hurley, B., O'Sullivan, B., Allouche, D., Katsirelos, G., Schiex, T., Zytnicki, M., de Givry, S.: Multi-Language Evaluation of Exact Solvers in Graphical Model Discrete Optimization. Constraints 21(3), 413–434 (2016)
38. Janin, J., Wodak, S., Levitt, M., Maigret, B.: Conformation of amino acid side-chains in proteins. Journal of molecular biology 125(3), 357–386 (1978)
39. Khalil, A.S., Collins, J.J.: Synthetic biology: applications come of age. Nature Reviews Genetics 11(5), 367–379 (2010)
40. Khare, S.D., Kipnis, Y., Greisen, P., Takeuchi, R., Ashani, Y., Goldsmith, M., Song, Y., Gallaher, J.L., Silman, I., Leader, H., Sussman, J.L., Stoddard, B.L., Tawfik, D.S., Baker, D.: Computational redesign of a mononuclear zinc metalloenzyme for organophosphate hydrolysis. Nature chemical biology 8(3), 294–300 (2012)
41. Khoury, G.A., Smadbeck, J., Kieslich, C.A., Floudas, C.A.: Protein folding and de novo protein design for biotechnological applications. Trends in biotechnology 32(2), 99–109 (2014)
42. Kingsford, C.L., Chazelle, B., Singh, M.: Solving and analyzing side-chain positioning problems using linear and integer programming. Bioinformatics 21(7), 1028–36 (2005)
43. Kuhlman, B., Baker, D.: Native protein sequences are close to optimal for their structures. Proceedings of the National Academy of Sciences of the United States of America 97(19), 10383–8 (2000)
44. Larrosa, J.: On arc and node consistency in weighted CSP. In: Proc. of AAAI'02, pp. 48–53. Edmonton, CA (2002)
45. Larrosa, J., de Givry, S., Heras, F., Zytnicki, M.: Existential arc consistency: getting closer to full arc consistency in weighted CSPs. In: Proc. of IJCAI'05, pp. 84–89. Edinburgh, Scotland (2005)
46. Larrosa, J., Schiex, T.: Solving weighted CSP by maintaining arc consistency. Artificial Intelligence 159(1-2), 1–26 (2004)
47. Leach, A.R., Lemon, A.P.: Exploring the conformational space of protein side chains using dead-end elimination and the A* algorithm. Proteins 33(2), 227–39 (1998)
48. Leaver-Fay, A., Tyka, M., Lewis, S.M., Lange, O.F., Thompson, J., Jacak, R., Kaufman, K., Renfrew, P.D., Smith, C.A., Sheffler, W., Davis, I.W., Cooper, S., Treuille, A., Mandell, D.J., Richter, F., Ban, Y.E.A., Fleishman, S.J., Corn, J.E., Kim, D.E., Lyskov, S., Berrondo, M., Mentzer, S., Popović, Z., Havranek, J.J., Karanicolas, J., Das, R., Meiler, J., Kortemme, T., Gray, J.J., Kuhlman, B., Baker, D., Bradley, P.: Rosetta3: an object-oriented software suite for the simulation and design of macromolecules. Methods Enzymol. 487, 545–574 (2011)

49. Lecoutre, C., Roussel, O., Dehani, D.: WCSP integration of soft neighborhood substitutability. In: Proc. of CP'12, pp. 406–421. Quebec City, Canada (2012)
50. Lewis, J.C., Bastian, S., Bennett, C.S., Fu, Y., Mitsuda, Y., Chen, M.M., Greenberg, W.A., Wong, C.H., Arnold, F.H.: Chemoenzymatic elaboration of monosaccharides using engineered cytochrome p450bm3 demethylases. Proceedings of the National Academy of Sciences 106(39), 16550–16555 (2009)
51. Looger, L.L., Hellinga, H.W.: Generalized dead-end elimination algorithms make large-scale protein side-chain structure prediction tractable: implications for protein design and structural genomics. Journal of molecular biology 307(1), 429–45 (2001)
52. Loudni, S., Boizumault, P.: Solving constraint optimization problems in anytime contexts. In: Proc. of IJCAI'03, pp. 251–256. Acapulco, Mexico (2003)
53. Lovell, S.C., Word, J.M., Richardson, J.S., Richardson, D.C.: The penultimate rotamer library. Proteins 40(3), 389–408 (2000)
54. Martin, V.J., Pitera, D.J., Withers, S.T., Newman, J.D., Keasling, J.D.: Engineering a mevalonate pathway in Escherichia coli for production of terpenoids. Nature biotechnology 21(7), 796–802 (2003)
55. Mladenović, N., Hansen, P.: Variable Neighborhood Search. Comput. Oper. Res. 24(11), 1097–1100 (1997)
56. Nestl, B.M., Nebel, B.A., Hauer, B.: Recent progress in industrial biocatalysis. Current Opinion in Chemical Biology 15(2), 187–193 (2011)
57. Noguchi, H., Addy, C., Simoncini, D., Wouters, S., Mylemans, B., Van Meervelt, L., Schiex, T., Zhang, K.Y., Tame, J.R., Voet, A.R.: Computational design of symmetrical eight-bladed β-propeller proteins. IUCrJ 6(1) (2019)
58. O'Meara, M.J., Leaver-Fay, A., Tyka, M., Stein, A., Houlihan, K., DiMaio, F., Bradley, P., Kortemme, T., Baker, D., Snoeyink, J., Kuhlman, B.: A combined covalent-electrostatic model of hydrogen bonding improves structure prediction with rosetta. J. Chem. Theory Comput. 11(2), 609–622 (2015)
59. Ouali, A., Allouche, D., de Givry, S., Loudni, S., Lebbah, Y., Eckhardt, F., Loukil, L.: Iterative Decomposition Guided Variable Neighborhood Search for Graphical Model Energy Minimization. In: Proc. of UAI'17, pp. 550–559. Sydney, Australia (2017)
60. Ouali, A., Allouche, D., de Givry, S., Loudni, S., Lebbah, Y., Loukil, L., Boizumault, P.: Variable neighborhood search for graphical model energy minimization. Artificial Intelligence (2019). https://doi.org/10.1016/j.artint.2019.103194
61. Pabo, C.: Molecular technology. Designing proteins and peptides. Nature 301(5897), 200 (1983)
62. Peisajovich, S.G., Tawfik, D.S.: Protein engineers turned evolutionists. Nature methods 4(12), 991–4 (2007)
63. Pierce, N., Spriet, J., Desmet, J., Mayo, S.: Conformational splitting: A more powerful criterion for dead-end elimination. Journal of computational chemistry 21(11), 999–1009 (2000)
64. Pierce, N.A., Winfree, E.: Protein design is NP-hard. Protein engineering 15(10), 779–82 (2002)
65. Pleiss, J.: Protein design in metabolic engineering and synthetic biology. Current opinion in biotechnology 22(5), 611–7 (2011)
66. Raha, K., Wollacott, A.M., Italia, M.J., Desjarlais, J.R.: Prediction of amino acid sequence from structure. Protein science 9(6), 1106–19 (2000)
67. Sánchez, M., de Givry, S., Schiex, T.: Mendelian error detection in complex pedigrees using weighted constraint satisfaction techniques. Constraints 13(1-2), 130–154 (2008)
68. Schiex, T.: Arc consistency for soft constraints. In: Proc. of CP'00, pp. 411–424. Singapore (2000)
69. Simoncini, D., Allouche, D., de Givry, S., Delmas, C., Barbe, S., Schiex, T.: Guaranteed discrete energy optimization on large protein design problems. Journal of Chemical Theory and Computation 11(12), 5980–5989 (2015)
70. Swain, M., Kemp, G.: A CLP approach to the protein side-chain placement problem. In: Principles and Practice of Constraint Programming–CP 2001, pp. 479–493. Springer (2001)

71. Terrioux, C., Jégou, P.: Hybrid backtracking bounded by tree-decomposition of constraint networks. Artificial Intelligence **146**(1), 43–75 (2003)
72. Traoré, S., Allouche, D., André, I., de Givry, S., Katsirelos, G., Schiex, T., Barbe, S.: A new framework for computational protein design through cost function network optimization. Bioinformatics **29**(17), 2129–2136 (2013)
73. Traoré, S., Roberts, K.E., Allouche, D., Donald, B.R., André, I., Schiex, T., Barbe, S.: Fast search algorithms for computational protein design. Journal of computational chemistry **37**(12), 1048–1058 (2016)
74. Verges, A., Cambon, E., Barbe, S., Salamone, S., Le Guen, Y., Moulis, C., Mulard, L.A., Remaud-Siméon, M., André, I.: Computer-aided engineering of a transglycosylase for the glucosylation of an unnatural disaccharide of relevance for bacterial antigen synthesis. ACS Catalysis **5**(2), 1186–1198 (2015)
75. Voigt, C.A., Gordon, D.B., Mayo, S.L.: Trading accuracy for speed: A quantitative comparison of search algorithms in protein sequence design. Journal of molecular biology **299**(3), 789–803 (2000)

Chapter 5
Modeling and Simulation in Dialysis Center of Hedi Chaker Hospital

Ichraf Jridi, Badreddine Jerbi, and Hichem Kamoun

Abstract Research and studies on modeling and simulation attached great importance to the health care field which is a rich field of investigation. The implementation of these two approaches has become more prevalent in order to support decision-making, to provide adequate information to physicians and managers, in order to predict and assist effective future planning. The main objective of this paper is to build a discrete event simulation (DES) model that includes the different stages that may go through an end-stage renal failure (ESRF) patient in a renal unit. These stages start by patient arrival and end by the death of the patient or by a successful kidney transplant. Switching a number of times between modalities of treatment is possible and presented in the model. Results show that the number of patients undergoing hemodialysis, peritoneal dialysis, or kidney transplant will continue to increase each year, which implies the increasing demand for treatment, required staff and materials.

5.1 Introduction

ESRF is a truly global public health problem. In 2011, the cost of management of patients on dialysis has surpassed 90 million dinars (37.000 euro) in Tunisia, nearly 5% of the overall health spending [6]. Therefore, ESRF constitutes a valuable socio-economic charge.

I. Jridi (✉)
College of Business Administration, Taibah University, Medinah, Kingdom of Saudi Arabia

B. Jerbi
College of Business and Economics, Qassim University, Al Malida, Kingdom of Saudi Arabia

H. Kamoun
Faculty of Economics and Management, University of Sfax, Sfax, Tunisia
e-mail: Hichem.Kamoun@fsegs.rnu.tn

© Springer Nature Switzerland AG 2021
M. Masmoudi et al. (eds.), *Operations Research and Simulation in Healthcare*, https://doi.org/10.1007/978-3-030-45223-0_5

ESRF patients are in a critical situation where there is currently no known cure for their condition and the damage done to their kidneys is unfortunately irreversible.

Indeed, the number of new patients arriving at the terminal stage is progressively increasing. In our country, few data pertaining to ESRF patients exist because of the lack of a national registry. Ben Maïz [3] has demonstrated that the number of ESRF patients went from 13 per million inhabitants in 1986 to 133 per million inhabitants in 2008. Hemodialysis (HD) was the most chosen technique of dialysis. It was undertaken by 96% of patients. Also, he has confirmed that a sparkling (or fuliginous) growth of HD was revealed from the year 1986. The number of patients undergoing HD increased from 97 in 1986 to 1374 in 2008.

From the first meeting with the renal unit managers, it was clearly seen that the characteristics of the redesign problem required a descriptive and adaptable modeling tool such as DES. The patient flow in the considered unit reflected a discrete set of events. Various studies in the health care field have utilized DES to model the operation of a system as a discrete sequence of events in time. Thus, the patient flow and the process in the renal unit under study are modeled using the DES tool. It provides an analysis of the current renal unit configuration, including robust estimates of the total needed resources.

The implementation and evaluation of health strategies related to the ESRF are necessary to improve prevention and disease management. Therefore, this research work is undertaken to model the chronic kidney failure path in the Sfax Governorate over the period of 11 years (from January 2000 to December 2010). Basically, it aims to design a valid model that will be useful for the improvement and decision-making process.

The remaining sections of this chapter are outlined as follows: Section 5.2 reviews the related literature. Section 5.3 contains a brief presentation of the renal unit at Hedi Chaker Hospital (HCH). Section 5.4 presents the details on how the renal unit simulation model was designed and built. Section 5.5 discusses and analyzes the obtained results. Finally, Sect. 5.6 provides concluding comments.

5.2   Historical Perspectives

Previous studies have revealed that simulation is the prominent method used for planning and resource utilization projects in health care [5, 14].

In 2003, the research of Fone et al. [11] has identified that the use of simulation had increased substantially since 2000. They have reported a systematic review of the literature on health care simulation.

The benefits obtained from the application of simulation modeling in health care are discussed in [16]. In fact, the authors have proposed seven differentiation axes, namely society's view and utopia (illusions), political influence and control, medical practitioners, health care support staff, health care managers, and patient's fear of death.

DES is an operations research (OR) modeling and analysis methodology that allows end-users (such as hospital administrators or clinic managers) to evaluate the efficiency of the existing health care delivery systems, by asking "what if?" questions, and to design their operations. DES can also be used as a forecasting tool to assess the potential impact of changes on patient flow to examine asset allocation needs (as in staffing levels or in physical capacity) and/or to investigate the complex relationships among different system variables (such as the rate of patient arrivals or the rate of patient service delivery). Such information permits health care administrators and analysts first to identify management alternatives that can be used, second to reconfigure existing health care systems, then to improve system performance or design, and finally to plan new systems without altering the existing system [12].

In a comparative study between DES and system dynamics, Brailsford and Hilton [4] have argued that DES usually considers the contexts to be simulated as queuing networks because individual entities pass through a series of discrete events one by one at discrete intervals. Between these intervals, entities have to wait in queues owing to the constrained availability of resources.

A systematic review conducted by Zhang [26] has presented the flexibility and new challenges perused by applying the DES tool to deal with a wide range of issues facing health care managers and shareholders. A survey of 211 papers was realized and the rapidly growing popularity of DES in the health care sector could be reflected by the substantially and constantly increasing number of publications, particularly after 2010. This may reveal that DES models are becoming increasingly involved in health care management.

The improvement of the efficiency of health care operations represents a challenging task because they are shaping the best way to organize the multiple resources required for the care delivery. Determining how to allocate and schedule these resources can rarely be performed by applying a simple formula [24]. Instead, managers and clinicians think of these decisions as systems consisting of complex relationships among interacting variables. DES is very useful for modeling these types of complex systems. Its use for these purposes is different from that for modeling analytic decisions since the analysis of operational systems generally requires a comparison of alternative systems to obtain the desired information [13].

Combining DES with health care has offered clinicians and managers the opportunity to evaluate possible improvement scenarios. A total of 117 papers have reviewed the application of DES in health care [7]. They demonstrated that very few studies have addressed the issue of complex integrated systems, and most of the studies focused on individual units within multi-facility clinics due to the extensive data requirement.

DES models that have been used to analyze health care delivery systems have primarily focused in two areas:

- Optimization and analysis of patient flow and
- Allocation of assets to improve the delivery of services.

The first area considers patient flow through hospitals and clinics, with the primary objective of identifying efficiencies that can be realized to improve patient through-put, reduce patient waiting times, and improve medical staff utilization. The second area considers the number of beds and medical practitioners requirements necessary to provide efficient and effective health care services.

Several benefits can be gained from conducting a DES study. Indeed, the procedure and methodology of applying DES, often for the first time, require decision-makers and managers to work closely with the simulation analyst to provide details of the system. As a result, the manager is likely to gain a new perspective on the relationships between the available resources and the quality of health care services offered by the system. The study of Rakich et al. [20] pertaining to the effects of DES in management development concludes that conducting a sim-ulation investigation not only develops a manager's decision-making skills, but also forces them to recognize the implications of system changes. Moreover, as proven by Wilson [25], when managers develop their own DES models, implementation occurs much more frequently. Finally, Lowery [18] has revealed that even if the implementation fails, benefits such as identifying unexpected problems unrelated to the original problem may emerge.

Abundant applications have been built using simulation in health care, such as those reported for ambulatory surgery centers, clinic laboratories, pharmacy staffing, emergency departments, cancer treatment centers, and renal replacement therapy centers or units [11].

As Zhang reported in his survey [26] the applications of DES in health care are divided into four classes, 65% of which represent health and care systems operation (HCSO), 28% correspond to disease progression modeling (DPM), 5% embody screening modeling (SM), and 2% stand for health behavior modeling (HBM). More than 68% of applications concerned to HCSO by DES were focused on specific problems within individual units.

In 2006, Davies [8] has noted that the ESRF is a fatal disease if it is not treated by a renal replacement therapy (RRT). For that she has used simulation modeling to determine the resource requirements for treating the ESRF. A comparison between the national level and the local level in the United Kingdom (UK) has occurred. Unlike the UK, where the estimation is a hard task, the increasing demand by new patients may be estimated with reasonable confidence by past rates of increase and current practices in other European countries.

Simulation has proved valuable in determining the need for local renal services at the district level. Davies and Roderick [9] have improved the use of DES to explore the overall demand for treatment in the UK. They have concluded that the number of patients needing treatment continues to increase in the future automatically accompanying costs raise.

A simulation model was developed in [22] to predict the future demand for RRT in England. The authors have noticed that the predicted number of new patients undergoing RRT has risen substantially over the decade by an average of 2.6 %. They have provided a range of estimates based on different assumptions about the unmet need because the demand estimates are uncertain.

Simulation plays an important role especially in designing or instituting health care facilities. Rohleder et al.[23] have used simulation to evaluate the optimum amount of laboratory facilities to serve the region. They have argued that it is downsizing, but is still providing the essential services.

Thus, DES is not only about data collection and output analysis, it also involves learning the complexity of the system and designing a valid model that is useful for training and decision-making [10].

Banks and Chwif's have outlined the key steps necessary to undertake a successful simulation study. To describe any process they have proposed using seven categories: data collection, building the model, verification and validation, analysis, simulation graphics, managing the simulation process, and human factors familiarity and abilities [2].

To model the ESRF patient flow and path in the renal unit of HCH in the region of Sfax, Tunisia, only data collection, building the DES model, verification and validation, analysis, and managing the simulation process are of special interest in the current research.

The following section provides a brief presentation of the renal unit under study.

5.3 Characteristics of the Renal Unit in HCH

A renal unit may be within a hospital or in a separate building. HCH is a university hospital that provides a clinical education and training to future and current doctors, nurses, and other health professionals. In addition it is delivering the medical care to patients. The renal unit is located within the hospital, not in a separate building. It is closely linked with the nephrology department. We found three inter-related sub-units in the renal unit of HCH. First, the HD unit that provides different treatments for HD patients. Second, the peritoneal dialysis (PD) unit that provides facilities and necessary oversights. Finally, the transplant unit where kidney transplants are carried out.

The HD's staff cares for HD patients who usually visit the HD unit three times a week for dialysis. Nurses at the HD unit prepare the equipment, insert needles, and supervise sessions. HD and PD units encourage patients to participate in their own treatment.

The PD's staff plans and monitors the treatment and care of PD patients. Patients come to the PD unit for their regular appointments (once after every 2 months), and some are trained there before starting PD therapy at home.

The transplant unit represents a part of the main nephrology ward too.

Generally, renal units are considering patients with the end-stage kidney failure as outpatients for the reason that they come to the unit for their appointments but are not required to get hospitalized. However, nephrology wards are existing in HCH renal unit, in which patients can stay for special observations or treatments such as scans, X-rays, medication, or minor surgery to insert a dialysis catheter or to create a fistula.

The practitioners' staff of the HCH renal unit are composed of two workers, 17 nurses—14 for HD unit, 2 for PD unit, and one administrative nurse (for the office)—a supervisor, a machine technician, a social worker, an associate professor doctor, a specialist doctor, a resident, and an internal medicine.

As equipment, there exist only 27 machines for HD treatment, out of which 20 machines are in work, 4 are in reserve and 3 are put in case of acute dialysis.

5.4 Simulation Model Development

Building a simulation model is a cyclical and developmental process as presented in Fig. 5.1.

Depending on the explanation and terminology found in the most used books of simulation, Robinson defines simulation as:

> Experimentation with a simplified imitation (on a computer) of an operations system as it progresses through time, for the purpose of better understanding and/or improving that system [21].

And Law states that:

> In a simulation we use a computer to evaluate a model numerically, and data are gathered in order to estimate the desired true characteristics of the model [17].

In general, to successfully execute a simulation model the following steps must be followed:

- Check out the real-world system and gather the needed data for building the simulation model;
- Abstract this real-world system and data and develop the simulation model that accomplished the study objectives;

Fig. 5.1 Simulation modeling cycles

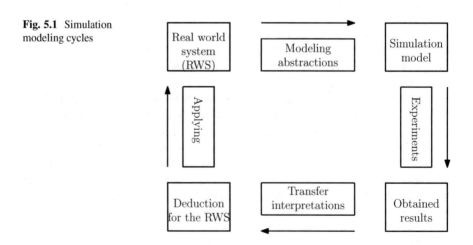

- Then, running experiments, i.e., execute simulation runs for the designed model;
- The next step will be interpreting the results and data that the simulation runs produced;
- Finally, management will use the results as a base to make decisions about optimizing the real system.

Patients flow is considered as the movement of patients through a set of locations in health care settings. Marcia et al. [19] have noted that the characteristics of patient flow represent the basic elements and assumptions in any simulated model. Six characteristics are recorded:

1. Extended waiting time and lists due to various and complex operations
2. Uncertainty and apparent chaos are common
3. Every patient is unique
4. The length of stay is proportional and differs from a patient to another
5. The incidence of complications
6. Emergency admissions

In this section, we first give a description of the process considered to build the DES model. Second, we provide a detailed description of the DES model. Next, we present and define the software used. Finally, verification and validation are performed.

5.4.1 Description of the Processes Considered to Build the DES Model

The process begins with the arrival of the patient to the renal unit and ends with his/her release either by being admitted to a successful transplant or dying. Upon arrival, the ESRF patient is registered and put on the transplant waiting list (TWL).

The available treatment modalities for a complete kidney failure are to replace its lost functions by the dialysis techniques or by the transplantation. Dialysis is an artificial process of filtering wastes and removing fluid from the body. It contains two choices of treatment:

- Hemodialysis (HD) or
- Peritoneal dialysis (PD).

During HD, the patient blood is cleaned using an artificial membrane. Blood travels to a dialysis machine where it goes through a special filter called a dialyzer before it returns to the body. The dialyzer removes wastes and tries to balance water, minerals, and chemicals in the blood. Only a small portion of the patient blood is out of his/her body at any time. The schedule for HD is three sessions a week. The model simulates conventional thrice-weekly HD sessions of 3.5 to 4.5 h duration, with an average of 4 h. The simulated facility has 24 HD machines and the medical staff operate on a 4-h double-shift schedule for six days a week.

PD is done independently at home, every day in two forms:

- Continuous Ambulatory Peritoneal Dialysis (CAPD): a technique that continues 24h/24h, fully manual.
- Automated Peritoneal Dialysis (APD): usually done at night (on average 8–10 h) using a small machine (cycler).

In Tunisia, we have very few patients on PD (with the rate of 4% from the total number of ESRF patients) because the social environment does not facilitate the use of PD as a viable renal replacement therapy option.

During the course of being treated for ESRF, patients may switch a number of times between the different alternatives of renal replacement therapy. For example, a given patient might move from PD to transplantation and, after transplant failure, to HD and perhaps to a second transplant: it is *the integrated care concept*.

The choice between HD and PD depends on many factors, including personal preference, health and medical suitability, lifestyle, availability of resources, and so on.

The patient health team/staff discuss the advantages and drawbacks of each type with both patient and his/her family. It may be possible to change between dialysis alternatives (HD or PD) if the chosen modality is not suitable for the patient. However, sometimes the choice is restricted. To have HD, it is necessary to have a good access to patient bloodstream (arteriovenous fistula), which can be an issue if he/she has diabetes. If the patient has heart problems, the major changes in blood pressure and waste levels caused by HD might engender a problem, in this case PD is preferred. PD may not be possible if the patient has major abdominal surgery causing scarring.

Kidney for transplant comes from living or deceased donors. The person getting the kidney is the recipient and the person giving the kidney is the donor. In Tunisia, living donors must be relatives such as father, mother, sister, brother, or any other member of the family. Deceased donors are people who previously have decided to donate their organs after death or who did not decide to donate in their lives and after their death their families accept the donation. Not everyone is eligible for a kidney transplant and the waiting list for suitable donor kidneys is usually overloaded. It is worthwhile to mention that a transplanted kidney generally functions like a normal kidney. Certainly, a successful transplant provides much more efficient kidney function than dialysis. Patients feel better and gain more energy.

The proposed simulation model describes the movement of patients through a series of events which are assumed to happen at discrete points in time. It does not consider putting a patient on TWL as an event because, in Tunisia, any patient reaches the ESRF is automatically registered in the TWL from his/her first coming to the renal unit. The patient will be removed from the TWL if he/she grafts a kidney and the surgery succeeds or when he/she dies. Actually, death is the only unexpected one-way out of the model.

Figure 5.2 includes a flowchart that indicates the possible movements of patients between the different states. The process flow in the renal unit under study was

Fig. 5.2 Generic flowchart
of ESRF patient flow
represented by the simulation
model

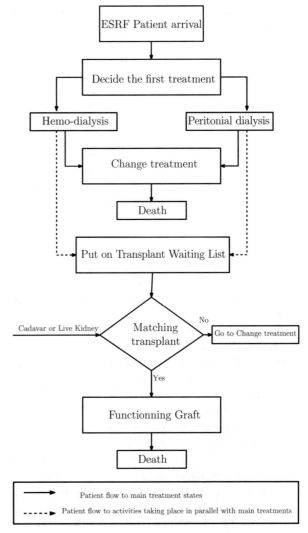

observed over a long period of time (11 years). The flowchart was prepared after
many alterations and subject matter experts' consultations.

5.4.2   DES Model

A DES model is not defined in terms of mathematical equations series, but rather
in terms of the physical movement of a transaction (e.g., patient, lab test, HD unit,
etc.) over time through different facilities and resources.

The movement of the ESRF patient, upon his/her arrival, within the renal unit of HCH represents a programming module written in SIMAN language.

Figure 5.4 gives a simple though comprehensive and descriptive idea of the work-flow. The logic was constructed and tested. Verification is defined as assuring that the computer program runs and its logic is robust. Validation means guaranteeing that the obtained results are in accordance with reality. Verification and validation are performed until the results make sense to the decision-maker.

For instance, *the create block* made in Arena is a subroutine to generate an ESRF patient. Each *create block* has its own characteristics, namely the name of the block, the entity type (in this case, the category the patient belongs to), the average time between arrivals, the inter-arrival distribution, the time unit, the entities per arrival, the maximum arrivals, and the creation time of the first arrival. The first *create block* is named "ESRF arrival," whose entity type is ESRF. The time between arrivals has a value equal to 28, where days is the measurement unit. The number of entity (patient) per arrival is one and infinite number can be considered as maximum arrivals.

Upon arrival, the ESRF patient is registered and put on the TWL. If no kidney is available for transplantation, the patient will decide (with doctors) on the first treatment which is HD or PD according to different criteria especially his/her health state.

The entity is processed through a *decision block*. This module makes the decision-making processes possible in the system. It includes options to decide based on one or more conditions or probabilities. Conditions can be based on attributes, variable values, entity types, or expressions. Entity here has to be checked for attributes (HD, PD, or RT). If the patient attribute is HD, *a creation block* will be created and called "HD patient Arrival," then the patient has to go through the HD process. Else, if the patient attribute is PD, *a creation block* will be created and named "PD patient arrival," then the patient has to go through the PD process. Otherwise, if it is RT, *a creation block* will be created and named "RT patient arrival," then the patient has to go through the RT process.

Thus, each treatment modality is represented by a *process block*. The latter is a module that has a logic entry which can be considered as: (1) delay simply indicating that the process will occur with no resources constraints; (2) seize-delay indicating that a resource (or resources) will be allocated in this module and a delay will occur, but that the resource(s) release will occur at a later time; (3) seize-delay release meaning that the resource(s) will be attributed followed by a process delay, and then the allocated resource(s) will be released. In our case, the type delay logic is assigned since the patient has no wait to be treated.

The process block module is intended to be the main processing method in the simulation model. Options for seizing and releasing resource constraints are available, besides the option to use a "sub-model" and specify hierarchical user-defined logic. The process time is allocated to the entity and considered to be value added since the time is deemed necessary to the entity. Table 5.1 represents the process time for all *process blocks* used. For the renal unit under study, the option is standard, i.e., non sub-model. In other words, no further details of the process are

Table 5.1 The process time for all treatment modalities

	Unit	Minimum	Most likely	Maximum
Hemodialysis	Hour	3.5	4	4.5
Peritoneal dialysis	Min	30	60	90
Transplantation	Min	130	240	280

required. The logic is of the type delay. All patients are considered to have the same priority and stay with the same acuity level during their stay in the renal unit.

Each *process block* is linked to *a decide block* to decide whether changing the current treatment modality to another or not. A PD patient can move to HD or RT, else he/she keeps carrying on PD day after day or can die at any time. Similarly, an HD patient can move to PD or RT; otherwise, he/she keeps carrying on HD one day after 2 or can die as well. An RT patient can leave the system if he/she gets a successful kidney transplantation or dies. Else, he/she can return to use one of the two dialysis forms, because no patient can gain two successive transplantations. He/she must dialyze after a failed graft.

The assign module is generally used to give values to attributes and variables. Six assign modules are used in case of changing the treatment modality. For example, the entity type will be modified from HD to PD if the patient changes dialyzing from HD to PD.

The discharge is expressed by *the dispose module*. It is intended to be the ending point for entities in simulation model. It represents the module that models the departure of entities from the system. It requires no parameters and simply removes an entity from the model destroying the contents of its attributes in the process. In our model, entities end with "Death" or "Success of transplant."

Data are collected through direct observations in the HCH hospital, particularly in the renal unit. The data includes arrivals of patients each year, types of patients (HD, PD, or RT), hospital resources (e.g., nurses, doctors, and the number of HD machines, etc.), the percentage of patients treated with the different modalities, processing times of each modality.

In this context, we used ESRF patients visits data from January 2000 to December 2010 as presented in Table 5.2.

For the simulation model, patient arrival times are used to create an exponential distribution of patient inter-arrival time, which is the time between consecutive patient arrivals. The exponential distribution is the standard distribution to use for fitting inter-arrival times.

Based on the collected data related to the ESRF patients arrival each year, it is to be noted that their number is on an upward trend especially in the last years from 2006 to 2010 as shown in Fig. 5.3.

The DES model also simulates interactions between the three available treatment modalities. The transfer frequencies are as follows: 95% from the ESRF patients go through hemodialysis, 4% undergo peritoneal dialysis, and 1% will reach the renal transplant.

Table 5.2 Number of patients undergoing HD in the renal unit of HCH

Year		2000	2001	2002	2003	2004	2005	2006	2007	2008	2009	2010
Number of HD patients	Semester 1	73	81	70	86	72	71	82	66	78	95	76
	Semester 2	71	51	76	69	84	65	82	74	83	88	78

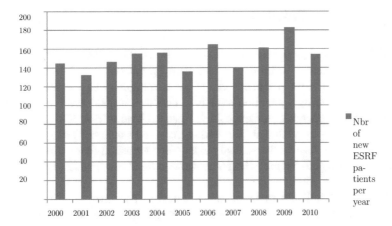

Fig. 5.3 ESRF patients arrival per year from 2000 to 2010

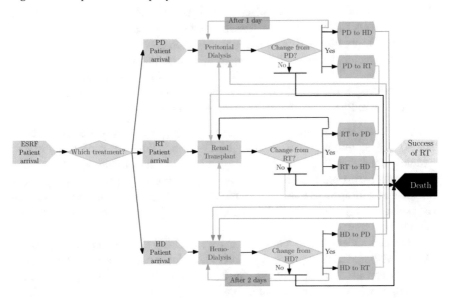

Fig. 5.4 Process chart

In the HD unit, nurses work in two shifts, the first of which starts from 07:30 am to 11:30 am and oversee the first session of HD. As for the second shift, the second team starts processing from 12:30 am to 16:30 pm. They manage the second session in the day.

Collected data and inputs are all handled by using the input analyzer from the Arena software which will be presented in the next section (Fig. 5.4).

5.4.3 Software Selection

To build and run the proposed simulation model we used the Arena version 12 (from Rockwell Automation Technologies Inc., USA) simulation Software package. We utilize the free but limited student version due to economic constraints.

Besides describing simulation models in health care that have been developed in the past two decades, Almagooshi [1] has reviewed 25 studies according to the simulation types (included DES and agent based simulation) and the software platforms used to develop the studied model. She focused on four platforms: Arena, Awesim, Matlab, and Simul8. Her study shows a trend of using Arena in DES was spotted.

The construction of simulation models with Arena involves using modeling shapes, called modules, from the basic process panel. These modules are used as the building blocks. There are two types of modules on a panel:

Flowchart modules: The user places flowchart modules in the modeling window and connects them to form a flowchart which makes the description of the model logic possible and easy. The most common flowchart modules are: Create, Process, Decide, Dispose, Batch, Separate, Assign, and Record.

Data modules: The user can edit these modules in the spreadsheet interface. These modules are not placed in the model window. The most common data modules are: Resource, Queue, Variable, Schedule, and Set.

The user builds an experiment model by placing modules (boxes of different shapes) that represent processes or logic. Connector lines are used to join these modules together and specify the flow of entities. While modules have specific actions relative to entities, flow, and timing, the precise representation of each module and entity relative to real-life objects is subject to the modeler. Statistical data, such as cycle time and WIP (work in process) levels, can be recorded and outputted as reports.

Arena is often employ to model health care facilities and problems. It represents a Windows-based platform that is popular and widely used due to its tremendous flexibility and ease of use. Models can be constructed without any programming knowledge due to its use of dialog boxes and graphical interface. It has been handled to model from manufacturing work cells to complex interactions in different organizations.

Furthermore, Arena's Input and Output Analyzers provide excellent tools to fit input probability distributions based on actual data and analyze output data using classical statistical measures. Animation of the model provides a visual representation for those without the technical understanding of a simulation modeling language.

For a better understanding, Kelton et al. [15] provide an excellent overview of simulation using Arena for beginners and experts.

5.4.4   Model Verification and Validation

Verification is the task of ensuring that the model behaves as intended according to the made modeling assumptions.

The simulation model has been verified by the renal unit's managers. They confirmed that the outputs of this model are reliable sources used for observing and monitoring the real situation.

This confirmation has been achieved by comparing the animation module created by Arena with what the renal unit's managers experience in reality.

They were validated against the number out of patients, 2 patients leave the system each year. This number is not statistically different from the actual scenario.

Moreover, the final results are validated using a different set of collected data which have not been used in the simulation process. The main reason for validation is to check whether or not the simulation model is an accurate representation of the real system.

5.4.5   Run and Set-Up Parameters

Simulation of the model was performed with 1000 replications (i.e., observations). This size was chosen to allow rules giving stable solutions. The duration of a replication was set to 365 days (1 year period) since it represents the length of session.

5.5   Results and Analysis

Results are automatically analyzed when the simulation running finished. A default report displayed at the end of the run. It summarizes the results across all replications.

During one year, there is an average of 3 patients dying while waiting for the renal transplantation. The maximum average of them is 8 patients. Normally, an efficient renal unit has a short patient waiting time for transplant.

The average accumulated time for treating an HD patient during one year is presented by 580 h. It is prior to the real time for HD sessions during one year which is equal to 576 h.

Patient undertaking PD performs his/her treatment at home. He/she has an appointment once every two months and the whole examinations have not exceeded 1.5 h. As an accumulated time during one year we obtain a total of 12 h.

The surgery of a renal transplantation lasts from 4 h to 5. As an average accumulated time we get 560 min.

Table 5.3 Number of patients remaining in the renal unit

	Average	Min average	Max average
Hemodialysis	8	3	17
Peritoneal dialysis	2	0	4
Transplant	1	0	1

The simulation predicted that as an average number 18 new ESRF patients will arrive to the renal unit of HCH each year (15 patients will be treated by HD, 2 patients will undergo PD, and only one patient will reach the renal transplantation). However, the average maximum number is 40 (32 new patients will start treating with HD, 5 patients will undergo PD, and 3 patient will be assigned to the renal transplantation).

Thanks to the WIP it is concluded that a number of patients are remaining in the system at the end of the replication running. Table 5.3 represents the WIP related to each treatment.

It is obvious that during the development of the HD *process module*, the need for HD treatment exceeds the capacity provided (using thrice weekly HD sessions), which is a situation that necessitates immediate intervention. Thus, in the renal unit of HCH, it will be necessary to allocate new resources (deployment of staff and equipment) if patients will not transfer to other available facilities. Knowing that one new machine can serve just 4 additional HD patients. Due to the obtained results a number of 4 new machines must be added to serve the 15 new HD patients.

Studying and simplifying the patient flow and path and incorporating the whole system information are essential. It is a basic and a promising task leading to improve the performance of health care setting operations.

The developed model serves as a reminder of the fact that the process of building DES models, especially in health care, not only help in future planning but it also represents a tool to clarify which processes really must take place in the organization under study. It enhances the idea that the DES can be a valuable tool for optimizing complex adaptive health care systems.

However, larger improvements are still expected. Regarding the progressively increasing of new ESRF patients each year, the renal unit will need to come up with new plans to increase the capacity especially of the HD unit (because 96% from ESRF patients will undergo HD). Also, various resources for high demand need to be invested.

5.6 Conclusion

Simulation plays a progressively important role in health care settings. Focusing on patient flow, it is a critical determinant of hospitals performance and patient outcome that has proven challenging task to optimize.

In this chapter, the flow and the path of ESRF patients in the renal unit at HCH in Sfax, Tunisia, are studied and simulated. The present model was formulated and supplied to the Arena software by careful programming and modules choice. It is verified and validated after many suggestions from the subject matter experts. All the guidelines to make a successful DES model were followed.

Different characteristics are reflected in the model. First it deals with individual patients. Second the model has to be a Non-Markovian since the future (e.g., survival with HD, PD, or transplant) of patients is conditionally dependent on their past (e.g., the patient is diabetic), given the present (i.e., the patient has reached ESRF). Third, patients belong to different discrete states. Finally the model examines interactions with the environment (e.g., availability of resources such as staff and HD machines and/or available treatment modalities).

It has important prospects, for local health decision-makers dealing with planning for the future. Because it may provide implications for national and regional health planners about the increasing number of patients. Also, it gives useful estimates of different resources required in the near future and represents a good medium for relating costs to estimates.

While the proposed model is validated in this study, key limitations should be considered. These included historical data from 2000 to 2010 may not have represented the renal unit and ESRF patient demand volume, staff availability, and other system characteristics at the current time, one hospital and an individual unit, the use of historical data that are specific to the renal unit of HCH limits the generalizable of the results to other renal units, small number of patients (due to the academic version of the simulation tool), lack of contextual data. It was more significant to make assumptions in the design of the DES model studied especially when input data were not available and when processes were not clearly defined in the real system. As with any model, assumptions can help to simplify the analysis. Moreover it can ensure and pressure the validity of the model if the analysis is imprecise. These limitations make the process redesign and the system operation with the stakeholders difficult

Future development will lead to an advanced configuration of the renal unit and will gather on the analysis. The current analysis could be extended to include queue analysis (e.g., length over time) and to allow direct and practical comparison of simulations run with different management decisions.

A deeper understanding of the patient flow at all levels of a health care system will facilitate future outlines which are corresponding to this critical factor in health care management.

Although modeling of specific units has a prevalent role in all the health and care systems operation areas, more scientific efforts have been analyzed to simulate more and different health care frameworks as hospitals or even the whole health service continuum that implies the rich potential of DES application to provide a clear and full picture of how health care systems act.

Despite the complexity of both the data requirements and the simulation program itself, this chapter demonstrates that simulation techniques can be successfully used in such a decision support system.

Acknowledgments The authors greatly appreciate the support of the staff and managers of the renal unit in HCH. This study could not have been completed without them. The authors also extend sincere thanks to Dr. Hichem Mahfoudh for the relevance of the information provided, his help, and his dedication to the study.

References

1. Almagooshi, S. (2015). Simulation modelling in healthcare: Challenges and trends. Procedia Manufacturing, vol 3, pp.301–307.
2. Banks, J. and Chwif, L. (2011). Warning about simulation. Journal of simulation, vol 5, pp.279–291.
3. Ben Maiz, H. (2010). Nephrology in Tunisia: From yesterday to now. Néphrologie and Thérapeutique, vol 6, pp.173–178
4. Brailsford, S.C. and Hilton, N.A. (2000). A comparison of discrete event simulation and system dynamics for modelling healthcare systems. Operational Research Applied to Health Services (ORAHS). Glasgow Caledonian University.
5. Brailsford, S.C., Harper, P.R., Patel, B. and Pitt, M.(2009). An analysis of the academic literature on simulation and modelling in health care. Journal of Simulation, vol 3, pp.130–140.
6. Chaabouni, Y., Yaich, S., Khedhiri, A., Zayen, M.A., Kharrat, M., Kammoun, K., Jarraya, F., Ben Hmida, M., Damak, J., Hachicha, J. (2018). Profil épidémiologique de l'insuffisance rénale chronique terminale dans la région de Sfax. doi:10.11604/pamj.2018.29.64.12159
7. Chemweno, P., Thijs, V., Pintelon, L. and Van Horenbeek, A. (2014). Discrete event simulation case study: Diagnostic path for stroke patients in a stroke unit. Simulation Modelling Practice and Theory, vol 48, pp.45–57.
8. Davies, R. (2006). Product Properties Essential for IEEE Computer Society. Proceeding of the 2006 Winter Simulation Conference, Monterey, CA, December, vol 3,:pp.473–477.
9. Davies, R. and Roderick, P. (1998). Planning resources for renal services throughout UK using simulation. European Journal of Operational Research, vol 105, pp.285–295.
10. Duguay, C., and Chetouane, F. (2007). Modeling and Improving Emergency Department Systems using Discrete Event Simulation. Simulation, vol 83, pp.311–319.
11. Fone, D., Hollinghurst, S., Temple, M., Round, A., Lester, N., Weightman, A., Roberts, K., Coyle, E., Bevan, G. and Palmer, S. (2003). Systematic review of the use and value of computer simulation modelling in population health and healthcare delivery. Journal of Public Health Medicine, vol 25, pp.325–335.
12. Jacobson, S.H., Hall, S.N. and Swisher, J.R. (2006). Discrete-Event Simulation of Health Care Systems. In: Hall R.W. (eds) Patient Flow: Reducing Delay in Healthcare Delivery. International Series in Operations Research and Management Science, vol 91. Springer, Boston, MA
13. Jerbi, B. and Kamoun, H. (2009). Using simulation and goal programming to reschedule emergency department doctors' shifts: case of a Tunisian hospital, Journal of Simulation, vol 3(4), pp.211–219, https://doi.org/10.1057/jos.2009.6.
14. Jun, J. B., Jacobson, S.H. and Swisher, J.R. (1999), Application of discrete event simulation in health care simulation : A survey. Journal of the Operational Research Society, vol 50, pp.109–123.
15. Kelton, W.D., Sadowski, R.P. and Sturrock, D.T. (2007). Simulation with Arena. 4th ed. New York: McGraw-Hill.
16. Kuljis, J., Ray, J.P. and Stergioulas, L.K. (2007). Can Healthcare benefit from modeling and simulation methods in the same way as business and manufacturing has?. Winter Simulation Conference.9–12 Dec. 2007 Washington, DC, USA.
17. Law, A.M. (2015). Simulation Modeling and Analysis (5th edn). McGraw-Hill.

18. Lowery, J.C. (1996). Introduction to simulation in health care. In Charnes J.M., Morrice D.M., Brunner D.T. and Swain J.J. (Eds.), Proceedings of the 1996 winter simulation conference (pp.78–84). Coronado, CA: Institute of Electrical and Electronics Engineers.
19. Marcia, W. and Bruce, S. (2005). Perfecting Patient Flow: Americas Safety Net Hospitals and Emergency Department Crowding. National Association of Public Hospitals and Health Systems, Washington, DC, 2005.
20. Rakich, J.S., Kuzdrall, P.J., Klafehn, K.A. and Krigline, A.G. (1991). Simulation in the hospital setting: Implications for managerial decision making and management development. Journal of Management Development, vol 10(4), pp.31–37.
21. Robinson, S. (2014). Simulation: The Practice of Model Development and Use (2nd edn). Palgrave Macmillan.
22. Roderick, P., Davies, R., Jones, C., Feest, T., Smith, S. and Farrington, K. (2004). Simulation model of renal replacement therapy: predicting future demand in England. Nephrology Dialysis Transplant, vol 19, pp.692–701.
23. Rohleder, T., Bischak, D. and Baskin, L. (2007). Modeling patient service centers with simulation and system dynamics. Health Care Management Science, vol 10, pp.1–12.
24. Rossetti, M., Trzinski, G. and Syverud, S. (1999). Emergency Department simulation and determination of optimal attending physician staffing schedules. In: Farrington PA, Nembhard HB, Sturrock DT and Evans GW (eds) Proceedings of the 1999 Winter Simulation Conference, pp 1532–1540.
25. Wilson, J.C.T. (1981). Implementation of computer simulation projects in health care. Journal of the Operational Research Society, vol 32, pp.825–832.
26. Zhang, X. (2018). Application of discrete event simulation in health care: a systematic review. BMC Health Services Research, vol 18 (687).

Chapter 6
Toward a Proactive and Reactive Simulation-Based Emergency Department Control System to Cope with Strain Situations

Sondes Chaabane and Farid Kadri

Abstract Emergency department is considered as one of complex services in hospital. This complexity is due to: care process complexity, interference with other services, influx of patients, high constraints, etc. Problems observed in EDs are mainly due to the organization, constraints, and changes in the care missions, as well as the mismanagement of process flows (patients, information, resources, etc.). This context makes the emergency department to face strain situations. However, health staff are neither prepared nor trained to cope with such context. Therefore, they need decision support systems adapted to anticipate and manage such situations. This paper proposes a proactive reactive simulation-based control system named ED-CS to design a tool that ED managers can use as a decision-making tool for anticipating and managing of strain situations, as well as planning the strategic deployment of corrective actions when such strains situations occur. This paper focuses on the decision-making process and system functions to be developed to implement corrective actions needed to anticipate and manage these strain situations. A case study is used to illustrate and show the potential benefits of ED-CS to deal with strain situations.

6.1 Introduction

Over the past decade, numerous reports and studies provide a complete picture of (comprehensive review) the state of hospital establishments in particular emergency departments (EDs) in material, financial, and moral crisis, [1–8]. Thus, these

S. Chaabane (✉)
CNRS, UMR 8201 - LAMIH-Laboratoire d'Automatique de Mécanique et d'Informatique Industrielles et Humaines, Univ. Polytechnique Hauts-de-France, Valenciennes, France
e-mail: sondes.chaabane@uphf.fr

F. Kadri
Analytics and Big Data Department, 1025-Aeroline PCS Agence, Sopra Steria Group, Colomiers, France

© Springer Nature Switzerland AG 2021
M. Masmoudi et al. (eds.), *Operations Research and Simulation in Healthcare*, https://doi.org/10.1007/978-3-030-45223-0_6

establishments are increasingly faced with difficulties to carry out their missions. Emergency departments (EDs) are an important component of healthcare systems because they provide immediate and essential medical care for patients. With the growing demand for emergency medical cares [9] and the reducing of number of EDs [6], the management of EDs has become more and more important, but they are also the most overcrowded component [10, 11]. Facing at a large number of patient visits but limited manpower, the ED has the very difficult task of trying to provide 24-h emergency services and offer a good quality service (minimizing patients' waiting times while not compromising the required attention for each patient) and ensuring that valuable resources (e.g., doctors' and nurses' time and treatment equipment) are well-utilized. The purpose of this paper is to present a proactive and reactive simulation-based system to aid EDs in anticipation and management of strain situations and for planning the deployment of resources (human and material) when such situations occur. It is organized in 8 sections. Section 6.2 defines and characterizes the strain situations, ED states, and strain indicators in an ED. Section 6.3 presents a decision support system in context of ED and strain situations. Section 6.4 characterizes and specifies the proposed proactive and reactive control system of an ED (CS-ED) in case of strain situations. Section 6.5 details the functions of proposed control system. Sections 6.6 and 6.7 are dedicated to the case of study and the results of the application of the principal functions of ED-CS to illustrate the interest of such functions. The last section provides concluding comments and future prospects.

6.2 Strain Situations in Emergency Department (ED)

Emergency department is dealing with mainly problems. It is a complex organization due to the complexity of care missions and the various process flows (patients, information, products, equipment, etc.). It must meet several constraints (structural, material, organizational, and financial). It is connected to other services in the hospital (operating theater, consultation, radiology, etc.) and outside the hospital (city doctor, pharmacy, special institutions, ageing homes, etc.). Different interferences between planned and unforeseen events and activities and interactions between different types of resources and skills (doctors, caregivers, nurses, administrators, managers, etc.) are observed in this system.

These problems cause the appearance of strain situations within the ED that affect patients and staff [12]. To handle these problems, ED managers must anticipate these strain situations by forecasting changes in ED demands (patient flows) and ED behavior, and, if necessary, to react quickly to the occurrence of these situations. Hence, EDs must incorporate in their operating mode the capacity to anticipate, react, and mobilize resources for satisfying both patients and personnel of EDs.

Most studies including emergency services have been addressing several aspects [13]. In [14–22], and authors are interested in improving the quality and performance of care in emergency departments. [23–28] focus their papers on reducing the waiting time and the residence time of emergency patients. Other works deal with resources management [29–34] and the hospital activities' management [35–37].

However, while the various works presented above are intended to improve the functioning of an ED, gaps can be observed. These gaps relate to both the definition and modeling of situations arising from the above issues and to avoidance and management strategies.

According to the interviews conducted with ED professionals and literature review, strain situations can be defined from various points of view and characterized by different factors. For example, strain situations can reflect the stress and fatigue of medical staff, or patients. From a patient flow viewpoint, a strain situation can be defined as an imbalance between the care demand and the capacity to provide care over a sufficiently long period [13]. This phenomenon can be related to overcrowding situation, but we maintain "strain" term to emphasize the impacts of such situations: stress, fatigue, dissatisfaction, violence ...

The main identified factors that may affect this equilibrium are [13, 38]:

- Patient flow and number of patients: seasonal epidemics (influenza, colds, gastroenteritis, bronchiolitis ...), aging population, crisis situations, accidents ...
- Care production capacity: capacity of medical staff (number and skills), availability of resources, internal and external transfer capacity ...

Based on the definition of a strain situation proposed above, the operation and/or behavior of an ED can be represented by two situations *(normal, strain)* and characterized by five main states *(normal, minor, moderate, severe, and critical)* as shown in Fig. 6.1 and detailed in [13]. Different measures are: V_e is the estimated

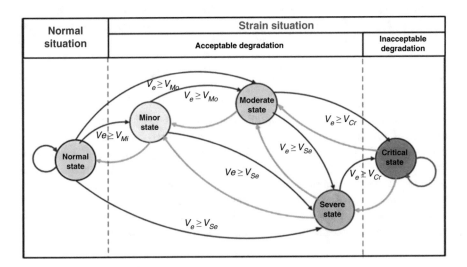

Fig. 6.1 Situations and transitions between states in an emergency department [38]

value of the measured parameter that characterizes the actual state of the ED; V_{Mi}, V_{Mo}, V_{Se}, and V_{Cr} are the equivalent threshold values that correspond the transition from one state to another according to the value of V_e.

The different states are detailed below:

- *Normal state*: This state corresponds to the current management based on planned actions and expected routine daily perturbations.
- *Degraded state (minor, moderate, and severe)*: The care flow is higher than the care production capacity. These three states are defined by:

  1. *Minor degraded state*: where the first threshold value V_{Mi} is exceeded ($V_e \geq V_{Mi}$);
  2. *Moderate degraded state*: where the threshold value V_{Mo} is exceeded ($V_e \geq V_{Mo}$);
  3. *Severe degraded state*: where the threshold V_{Se} ($V_e \geq V_{Se}$) is exceeded.

- *Critical state*: it is the worst case where the ED goes from severe state (acceptable degradation) to critical state (unacceptable degradation) where the threshold V_{Cr} ($V_e \geq V_{Cr}$) is exceeded. In this case, the emergency management must deploy important actions like "Plan Blanc" (White Plan) in order to return to an acceptable operating state.

Identifying the dynamic state of the ED and evaluating the threshold values imperatively require relevant indicators. One can find many types of strain indicators in the literature: waiting times, current number of patients present in the ED, length of stay in the ED … [39]. However, the characteristics of these indicators are not all accessible or usable. They must imperatively be established and validated by professionals of the ED to ensure their effectiveness to identify, quantify, and represent stress situations in an emergency service in different contexts. In reality, and according to the discussion with medical staff, the number of relevant indicators is about 5 or 7 with determined threshold value for each indicator. The dynamic state of the ED can be modeled by the raw data matrix $[SI]$:

$$[SI] = [SI_1, SI_2, \ldots . SI_i, \ldots , SI_n]^T, \tag{6.1}$$

where SI_i are strain indicators.

A SI_i can be a number, measured directly in the emergency department (e.g., number of hospitalization for more than 24 h) or calculated from measurements performed in the emergency department (e.g., the average length of stay based on the urgency in the last 24 h). ED managers need to design and implement approaches providing at the appropriate time and in the appropriate place information and specific knowledge. An effective support system to control the ED based on defined indicators and their threshold may be defined [40, 41]. In the next section decision context and decision-making process under strain situations will be presented.

6.3   Strain Situations in Emergency Department (ED)

Decision support system's design and implementation are attracting a great deal of interest from researchers in different domains. First a literature review in emergency department is presented. Second, the decision-making process under strain situations is exposed to highlight the ED manager's needs according this context.

6.3.1   Literature Review on Decision Support Systems in Emergency Department

In the case of the emergency department's activities, the literature review related to decision support systems (DSSs) and approaches is usually applied to improve the management of patient and ED resources.

[4] developed a decision system based on discrete event simulator of patient flow for predicting overcrowding in an ED. The inputs and outputs of the model are descriptions of six variables of each patient present in the ED. Validation of the model was performed using the real patient data of the ED. [42] presented a methodology that uses the combined simulation with optimization to determine the optimal number of doctors, nurses, and laboratory technicians needed to maximize patient throughput, reduce the patient waiting time, and assess the impact of various staff on the effectiveness of services. [43] proposed a simulation-based decision support system using a two-step procedure for the ED control in real time. In the first step, the ED manager detects the current state of the ED based on historical data and simulation: the data is sent to simulator and this later then provides the status of unobservable components. In the second stage, based on the current state, the simulator supports the control by predicting the future scenarios by estimating resource requirements. [44] proposed a simulation-based decision support system that provides a global perspective on the factors affecting the operation of the emergency department. The authors studied the causes of bottlenecks and inadequate distribution of resources, and a modeling approach has been developed through which the flow of patients is studied in relation to the medical staff. [45] proposed a decision support system for emergency departments based on agent-based techniques. Two distinct types of agents have been identified: the active agents representing medical staff, and the passive agents to represent the services. The model also includes the communication system and the environment in which agents move and interact. [36] presented a simulation-based decision support system for the effective exploitation of emergency departments. The main objective of the simulation is to optimize the ED performances. The optimization is performed to find the optimal configuration of staff (doctors, nurses, and triage staff) and admission staff. [46] designed a framework for an intelligent decision system to optimize practice's pathology test ordering by general practitioners.

It illustrates the processes of discovering practical and relevant knowledge from pathology request data generated and stored in a professional pathology company. It also investigates and understands the decision makers through a survey about their current practices in test ordering and their requirements for decision support and proposes an intelligent decision support framework as the decision strategy to support GPs in ordering pathology tests more effectively and appropriately. The developed framework may be used as an effective guidance for practitioners and theoretical understanding concerning intelligent decision support in a complex environment. Although many approaches and decision support systems have been utilized by emergency managers in the process of decision-making and improving the quality of care, only a few models and systems are relevant to effectively manage the strain situations [13, 47]. Simulation-based approaches are the most widely used tool and appear to be best suited to the problems and context of emergency services. Our contribution is to define the context of ED facing strain situations as defined in the previous section. Our aim is to present the decision-making process under strain situations and to deal with such situations to reduce their impact.

6.3.2   Decision-Making Process Under Strain Situations

The ED environment is dynamic and disrupted because of interferences between activities and no determinism of some elements that characterize the care activities (processing time, expectations, additional examinations). The patient's arrival at the ED is unexpected; each patient needs a treatment adapted to his or her condition that involves a specific path within the ED. There is no prior knowledge about patients requiring emergency treatment within a given period of time. The decision maker is faced with a "decisional situation" [48]. This concept reflects the fact that: the analysis of a decision must integrate the context in which it is taken [49]. The time available to take decisions varies according to the situation. For the ED manager, the emergency space-time is often measured in minutes, and the challenge (real or perceived) is a factor in this situational relativity. The difficulty lies in this intersection between the relative decrease in time available and the challenge. Furthermore, decisions in emergency contexts are complex and prone to errors. These factors make decision-making context of EDs under strain situations highly dynamic and complex.

In the case of emergency services, the situations encountered in the functioning of the ED are complex, and time constraints usually involve making decisions in a very limited time, especially in reactive management. Furthermore, available information is incomplete and/or imperfect. Note also that, given the context, the decision maker has limited information-processing capabilities. This, therefore, leads to choosing corrective actions and making decisions that are "satisfactory," but not optimal. This feature also tends to favor the use of simulations as support tools for decision-making.

In the emergency department context and according to the detected strain situation, three cases are identified:

- *Case 1*: detected strain situation is known, and the manager knows the suitable and corrective action. In this case the decision is made directly and an appropriate corrective action(s) that responds effectively to the detected situation is applied.
- *Case 2*: detected strain situation is similar to detected one in the past. In this case, the ED manager has the choice of applying the identified actions that are considered effective. Each corrective action identified and implemented for a given strain situation will be stored in the knowledge base for managing similar strain situations in the future.
- *Case 3*: detected strain situation is new, and the knowledge base does not contain any corrective actions corresponding to the actual situation. In this case, ED manager may propose a new corrective action and verify its effectiveness by using simulations and evaluations. If the proposed corrective action(s) are effective, he or she applies them directly in the system; if not, he or she proposes other actions and simulates their effects on the ED behavior, and so on. These actions are then stored in the knowledge base.

Figure 6.2 presents the decision-making process to apply corrective actions in the three cases presented above:

- *Identification of the ED state*: in this step the state of the ED is identified through the defined states *(normal, degraded, and critical)*.
- *Evaluation of the ED*: in this step strain indicators are used to evaluate the ED. The objective is to detect strain situation.
- *Choice of the best-suited corrective actions*: this step activates one of the three activities according to the identified strain situation presented above:
  - *If Case 1*: the ED manager applies an effective identified action directly.
  - *If Case 2*: the ED manager researches a similar situation and evaluates and validates the identified corrective actions before application.
  - *If Case 3*: the ED manager faces a new situation. In this case he or she must be able to propose a new corrective action and needs to forecast the behavior of the ED. He or she must evaluate and validate the identified action before application.
- *Application of actions*: it is the implementation of the corrective action in real system.

According to this context, ED manager needs a decision support system to help him or her in decision-making process described above. The overall objective in the implementation of decision support system to cope with strain situations is to

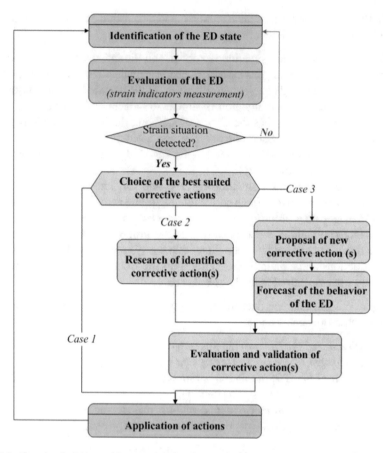

Fig. 6.2 Generic decision-making model for the choice of a corrective action under strain situations

improve proactive and reactive management of the ED by:

- Ensuring the development of relevant strain indicators to obtain the best representative picture of the status of the ED.
- Predicting short and/or medium term; and the detection of strain situations, based on the strain indicators.
- Assisting the implementation of effective and appropriate corrective actions for each potential occurrence of a strain situation and the evaluation of these actions.
- Capitalizing on the knowledge about ED behavior, to better cope with future strain situations (learning, experience, feedback, etc.) and promote training.

To respond to these needs, we present in the next section the proposed decision support system based on proactive and reactive control systems.

6.4   Proactive/Reactive Control System

As presented above in Fig. 6.2, ED managers must take decisions in two operating modes: proactive and reactive. In *proactive mode*, the ED manager must be able to detect possible strain situations, and the decision must be processed in a proactive context (proactive control). The proactive control covers the predictive control (predict an activity for which the decision is based on parameters that are estimated using either deterministic or probabilistic approaches) and the anticipation of disturbances. Based on various scenarios, he or she identifies the risks of not maintaining the ED under normal situation when disruptions occur. It concerns both the medium and short terms (month, week, day, and hour) and the long term (year). In *reactive mode*, the ED manager has also to react if a situation occurs, which must then be processed in a reactive context (reactive control). Reactive control occurs in short term or real-time, according to the occurrence of unanticipated events and/or disturbances. This control is thus made while the ED is functioning and without anticipation. It concerns the very short term (minute or hour). It is necessary when: (1) an unforeseeable and unexpected event occurs or (2) there are deviations, which may lead to a degraded or critical state of the ED.

In both modes, the decision maker must:

- *Identify the ED state* : in the case of both proactive and reactive controls.
- *Research corrective actions* in : (i) the medium term, if a strain tense situation may occur (*proactive control*); and (ii) in both the short and very short terms, if a strain situation occurs (*reactive control*).
- *Assess the impact* of this corrective action on the behavior of the ED.
- *Launch the corrective actions* if he or she estimates that these corrective actions are satisfactory or search for an alternative.

These two modes correspond to an operating process described in the next subsections.

6.4.1   Proactive/Reactive Operating Process

Table 6.1 presents the caption of different times and dates used in proactive and reactive modes.

The two operating modes are presented below:

1. **Case of proactive mode** is illustrated in Fig. 6.3.

    - At date t_{Fj}, a forecast is launched. If a strain situation is detected on date t_D, the ED manager has to seek a corrective action. A T_R time ($T_R = t_S - t_R$) is required for searching and choosing what action to activate. Depending on the type and severity of the disturbance, T_A ($T_A = T_P + T_L + T_{Rec}$) can be estimated based on the experience of the manager, or by a simulation.

Table 6.1 Captions of Figs. 6.3 and 6.4

	Indicators and thresholds
SI_i	Current value of the observed strain indicator
V_{Mi}	Threshold of minor degraded state
V_{Mo}	Threshold of moderate degraded state
V_{Se}	Threshold of severe degraded state
V_{Cr}	Threshold of critical state
	Dates
t_{Fj}	jth Launch date of forecasting
t_{SS}	Forecasted date of occurrence of a strain situation
t_C	Current date
t_O	Date when the disturbance occurs
t_D	Date when the disturbance is detected
t_R	Date of starting the research of corrective action begins
t_P	Date when the corrective action is prepared
t_L	Date when the corrective action is launched
t_I	Date when the corrective action takes effect within the ED
t_E	Date when the ED returns to normal state
	Times
T_D	Time required to detect disturbances
T_R	Time required to detect the time required to search corrective actions
T_P	Time required to set up corrective action
T_L	Time required to launch the corrective action
T_{Rec}	Time required for the ED to recover its normal state after the disturbance
T_A	Time required to activate corrective action

  Corrective action must be effective to make ED able to return to normal situation before t_{SS} ($t_E \leq t_{SS}$).

- Then if $T_A < t_E \check{~} t_R$, the manager can delay the activation of the corrective action according to a flexible margin ($t_P - t_S$). If not, $T_A \geq t_E \check{~} t_R$, and if no other action can be found, the corrective action must be activated immediately.

2. **Case of reactive mode** is illustrated in Fig. 6.4.

   In this case, corrective action has to be activated as soon as it is selected ($t_S = T_P$). Sometimes the disturbance is not detected immediately, but after the time T_D ($T_D = t_D - t_O$). The search for a corrective action then begins at $t_R = t_D$.

6.4.2 Proactive/Reactive Emergency Department Control System

To deal with strain situations, proactive and reactive modes must be combined. In this paper we propose a generic decision support system for an ED department

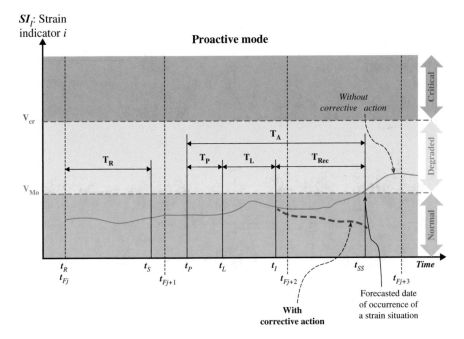

Fig. 6.3 Proactive mode

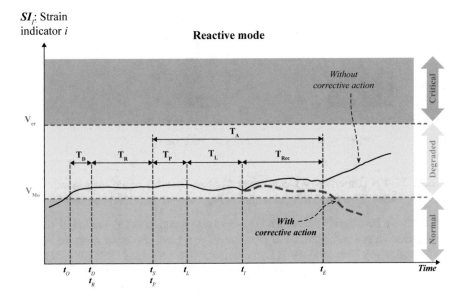

Fig. 6.4 Reactive mode

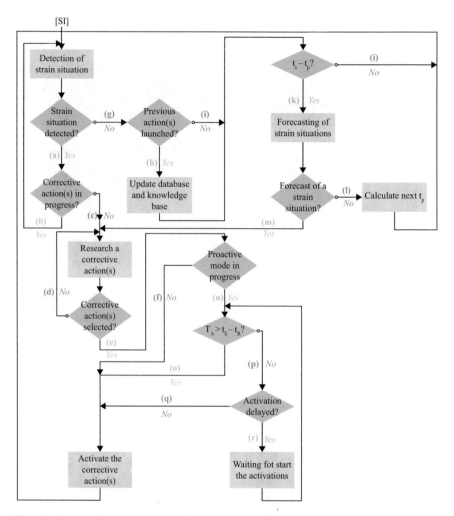

Fig. 6.5 Proactive/reactive emergency department control system

named ED-CS (Emergency Department Control System) based on proactive and reactive modes. The ED-CS is presented in Fig. 6.5 and described below:

- If a strain situation is detected by the detection function (*a*), the manager checks whether an action is already in progress. If yes, he or she waits until the end of the action (*b*). We assume that a single action or a combined action is enabled.
- If a strain situation is detected or if the previous action was not effective (a situation of tension is always present), a new action is required (*c*) until a new action is selected ((*d*) or (*e*)). In reactive mode, the action is immediately activated.

- If no situation is detected (g), the manager checks first if a previous action was launched. If it is the case, the database and the knowledge base are updated. Then, t_C and t_{Fj} are compared as if no previous action was launched. However, the manager checks if the current date corresponds to a forecast date ($t_C = t_{Fj}$). If yes, a forecast is launched (k). If not, the process returns to the detection phase (j).
- If a strain situation is detected (m), a corrective action is searched (as in the previous case). When an action is selected, in proactive mode (n), if $T_A > t_E \check{~} t_R$, the action is immediately activated (o). If $T_A \leq t_E \check{~} t_R$ (p), the manager decides either to delay the activation (r) or not (q) and then the action is activated.
- If no strain situation is forecasted (l), the next tFj+1 is calculated and the process returns to the detection as in (j).

The main functional requirements of ED-CS can then be identified as below:

- Provide the decision maker with the relevant strain indicators needed to work out a significant evaluation of the ED state.
- Ensure the proactive control based on the strain indicators, predicting short and/or medium term, and detecting the occurrence of strain situations.
- Ensure the reactive control to respond to undetected situations.
- Help to implement effective and appropriate corrective actions for each potential occurrence of strain situations and evaluate these actions.
- Store knowledge on the ED behavior, to better cope with future tense situations (learning, experience, feedback) and promote training.
- Provide models that allow experimenting with different corrective strategies in different conditions.
- Provide support for different categories of users, or groups of users.
- Provide assistance in staff training.

6.5   Functions of Proactive/Reactive ED-CS

To implement the ED-CS, the functional architecture must include the functions and tools presented below:

F1—Identification of the ED state
F2—Detection of strain situations
F3—Forecasting of strain situations
F4—Simulation of the ED behavior
F5—Data and knowledge storage function
F6—Evaluation and validation of corrective actions

6.5.1 F1: Identification of the ED State

The identification consists of constructing the strain indicators and dashboard. First, it needs to acquire the characteristic information (date of arrival of a patient, date of first consultation with a doctor, length of stay in the ED . . .) required to calculate the strain indicators. Second, the strain indicators defined by the emergency staff (e.g., length of stay in the ED, waiting time before seeing a nurse) are calculated. Third, the strain indicators' values are monitored and compared to the fixed thresholds defined by the emergency staff.

6.5.2 F2: Detection of Strain Situation

Monitoring approaches may be broadly classified into two main categories: data-based and model-based approaches [50–53]. Model-based anomaly detection approaches are based on the comparison of measured process variables with information retrieved from an explicit mathematical model, usually derived from a basic understanding of the inspected process under normal operating conditions [54, 55]. For more details and comparison between these two approaches, the reader can refer to the works of [53, 56–58].

Two combinations of methods can be used:

- Either, the times series analysis and the Statistical Process Control (SPC). SPC is widely used in the medical field for monitoring care in anesthesia, intensive care, risk management, and public health, including the control of infection rates, patient falls rates, death rates, and various types of waiting times [59].
- Or, Statistical Process Control and Principal Component Analysis (PCA). The PCA is the best known and the most widely used multivariate method in the monitoring process and anomaly detection [50, 52, 59].

6.5.3 F3: Forecasting of Strain Situations

The choice of forecasting functions can be achieved by taking into account the following constraints:

- All relevant data is relatively difficult to obtain in ED context. Therefore, we must choose methods that exploit the data generally available in the ED. Often, it is the historical evolution of certain characteristics of emergencies, including the number of entries to the emergency during a given period.
- The objective is primarily focused on proactive and reactive aspects; we have retained an extrapolative method to model and exploit these historical evolutions.
- The use and validation of these methods in EDs, the possibility of combination with other methods (detection methods), and their ease of implementation.

Therefore, the time-series models may be used for this purpose. In the field of healthcare systems, increasing attention has been accorded to time-series models to predict patient arrivals in hospital establishments [11, 60–62]. The literature review shows that time-series analysis has been largely applied in the hospital sector to forecast patient arrivals, length of stay, and for projecting the utilization of inpatient days. They are very useful as an initial significant analysis of patient flow because of their ease of use, implementation, and interpretation within the framework of modeling and forecasting of emergency department overcrowding, caused by the influx of patients [11].

6.5.4   F4: Simulation of the ED Behaviors

Simulation presents an important function of ED-CS. The simulation is a key tool for studying the behavior of complex dynamic systems [13, 44, 63]. It is an effective method, widely used to improve policies on operational, tactical and strategic decisions in EDs [13, 44, 64–70]. The simulation approach can incorporate randomness into the model, making it easy to examine many "what-if" scenarios (time- and category-dependent patient-arrival rates, different service-time distributions, and time-varying staffing levels). Moreover, compared with analytical approaches, simulations are less sensitive to model parameters [71]. In our case, the simulation involved in the ED behavior prediction phase is based on the decision-making process. It is used in proactive and reactive operating modes to research, test, and evaluate corrective actions. It is used for both priori and posteriori evaluations by: (1) predicting the evolution of the ED behavior; (2) visualizing the care process in the ED and evolution of tense situations based on associated indicators with each situation; and (3) enabling alternative tests (corrective actions, adaptation of the medical staff) before the implementation of the corrective actions, in order to analyze the simulated scenarios, thus improving decision-making and evaluating their effectiveness.

6.5.5   F5: Data and Knowledge Storage Function

The data storage concerns the information system of the ED. It contains: static data (structure, staff, number of rooms) and history of the ED activity (number of inputs, number of hospitalized patients, etc.). This system is completed by a knowledge-based system to store and update corrective actions and strain situations. It is based on a SSCase model including a set of information on the previous cases of strain situations to be memorized. This will help ED manager to find suitable corrective actions or to update the system with new actions. The proposed SSCase model is structured as follows:

$$SSCase = [C, O, SI, S_{ED}, V_{TH}, P_{CA}, A_{CA}, T_P, T_L, E], \qquad (6.2)$$

where:

1. C : in the framework of EDs, the search for corrective action(s) during the occurrence of a strain situation is strongly guided by the **context** in which the strain situation occurs. The C parameter defines three elements of the **context** in which the strain indicator is considered: (1) events context: specifies (if necessary) the event(s) or health threats (e.g., epidemics, seasonal flux, heat waves, cold waves ..., (2) situational context: indicates the ED state (e.g., degraded or critical) at the launch of one (or more) corrective actions, and (3) temporal context: defines the period or the time (hour of day, day of the week, and the period of year ...).
2. O: it is the **objective** of each strain indicator to be evaluated.
3. SI: it is the **strain indicator** as defined above.
4. S_{ED}: it presents the **ED state** (degraded or critical) after the occurrence of the strain situation.
5. V_{TH}: it presents the **threshold** values corresponding to the transition from one state to another for the concerned indicator following the occurrence of the perturbation.
6. P_{CA}: it is the **possible corrective action** associated with the strain indicator.
7. A_{CA}: it is the **applied corrective action**(s).
8. T_P: time required to **prepare** the corrective action.
9. T_L: time required to **launch** the corrective action.
10. E: indicates the **effectiveness** of the applied corrective action(s) expressed in terms of the required time of the ED to respond to a type of perturbation, for example, T_{REC} defined previously (Table 6.1).

6.5.6   F6: Evaluation and Validation Function

The evaluation function is based primarily on strain indicators and concerns the state of the ED and corrective actions. The ED manager can assess the condition of ED by monitoring the strain indicator values provided by the identification function. The evaluation process is performed according to the decision-making process by querying the knowledge base system and using the simulation (see Fig. 6.2). The corrective actions are: (i) either an identified corrective action, which the manager wants to test or (ii) a new corrective action proposed by the ED manager.

If, after review of the effectiveness of the proposed action(s), the ED remains in a degraded or critical state, then the manager may propose new corrective actions. A new simulation will be conducted to verify the effectiveness of these new actions. This process is repeated until the ED regains a normal operating condition. In this case, the effective corrective actions associated with the strain situation will be validated and implemented to the ED and stored in the knowledge base using the *SSCase* model.

6.6   Case Study

The Lille Regional Hospital Center (CHRU) provides care for the four million inhabitants of the Hauts-de-France, a region characterized by one of the largest population densities in France (9.4% inhabitants of the national population and the 3rd most populated in France). The pediatric emergency department (PED) in CHRU is open 24 h a day and receives 23,900 patients a year on average. Besides its internal capacity, the PED shares many resources, such as administrative patient registration, clinical laboratory, scanner, magnetic resonance imaging (MRI), X-rays, and blood bank, with other hospital departments. Based on the analysis of the questionnaires and interviews conducted with the PED medical staff, we established a dynamic model of the care process detailed in Fig. 6.6. Different activities are required depending on the type of patient and additional exam. In this paper, critical patients are not considered. They are treated immediately and do not wait.

According to states presented above in Fig. 6.1 and for this study, the behavior of the PED is characterized by two situations (normal and strain) and three main states (*normal, degraded, and critical*) with two thresholds: V_{Mo} (limit between *normal* and *degraded*) and V_{Cr} (limit between *degraded* and *critical*). The PED staff were involved (by means of questionnaires and interviews) in the selection and classification of the relevant strain indicators. The main strain indicators selected and validated are:

1. P_W: Primary waiting time is a waiting period between the end of admission by the hostess and the first medical examination.
2. T_2: waiting time between the end of the care by hostess and consultation by a nurse.
3. N_p: the current number of patients present in the PED at the arrival of a new patient.
4. PP: ratio of the number of patients present in the PED by the number of physicians.
5. Q_S: the ratio of the actual length of stay ($LOS(t)$) by the theoretical length of stay (LOS_{th}).

The strain indicators and their threshold values used in this study are defined in Table 6.2. The threshold values were defined and validated by the pediatric medical staff.

In this paper, we focus on corrective actions on human and material resources. In next section, we present the results obtained for several functions of the ED-CS in the case of proactive and reactive control of the studied PED. We used a detailed model of the PED to simulate its behavior and test the effectiveness of corrective actions.

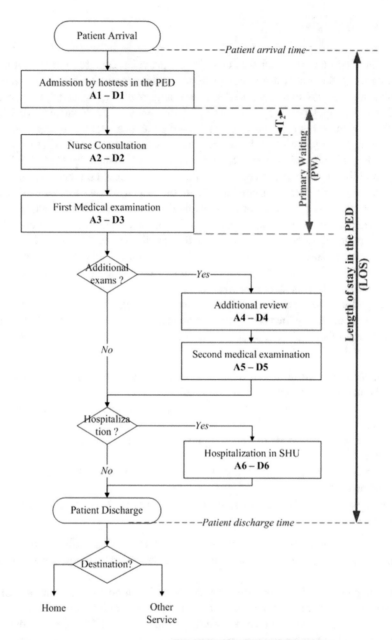

Fig. 6.6 Care process at the PED

Table 6.2 Selected strain indicators and their thresholds

	PED states		
Strain indicators	Normal	Degraded (V_{Mo})	Critical (V_{Cr})
P_W(min)	$P_W < 60$	$60 \leq P_W < 90$	$P_W \geq 90$
T_2(min)	$T_2 < 25$	$25 \leq T_2 < 50$	$T_2 \geq 50$
N_p(number)	$N_P < 12$	$12 \leq N_P < 20$	$N_P \geq 20$
PP	$PP < 4$	$4 \leq PP < 7$	$PP \geq 7$
Q_s	$Q_s < 0.8$	$0.8 \leq Q - s < 1.5$	$Q_s \geq 1.5$

6.7   Experiments and Results

We implement and experiment proactive and reactive control functions of the ED-CS for our case of study. Next subsections present some of the results.

6.7.1   Proactive Control Simulation Results

We evaluated the behavior of the ED, and the evolution of two strain indicators after the occurrence of a disturbance (increased arrivals of patients), both without the application of corrective actions and after applying corrective actions. The strain indicators selected by the ED staff for the experiments are: primary waiting time (P_W) and waiting time before nurse consultation (T_2). The daily evolution of the PW for a given Sunday of the winter period 2012 (considered as the busiest day of the winter period) was simulated, and the evolution is presented in Fig. 6.7. We used ARENA 14.00 simulation tool of Rockwell software. To consider the warm-up period, the simulations were run for three days and strain indicators measures are collected for the second day. We noted that patient arrivals at the PED increased considerably over 2 h (between 11 a.m. and 1 p.m.): 9 patients between 11:00 a.m. and 11:59 a.m., and 11 patients between 12:00 and 12:59 p.m. This had an almost immediate effect on the patient waiting time before consultation with a nurse, and a delayed effect on the first medical consultation (P_W).

To reduce the PW in the PED, additional resources (nurses and/or doctors) and material resources (consultation boxes) are added during the strain period. The proposed scenarios are presented in Table 6.3.

Figure 6.8 summarizes the results of the corrective actions for each proposed scenario. According to this figure; it can be observed that scenarios S2, S5, and S6 contribute to reduce the P_W to an acceptable level. It can also be noted that scenario S5 is economically more interesting than S6 because it only requires one additional resource, i.e., 1 doctor and 1 box.

To analyze the impact of corrective actions on the T_{Rec}, we applied corrective actions by using the two best scenarios S2 and S5 (identified previously) at $t_I = $ 12:30 with a launch time for corrective actions $T_L = 74$ min. In this case study, we

Fig. 6.7 Primary waiting time (P_W) in minutes without corrective actions (Scenario 0)

Table 6.3 Proposed
corrective actions

Scenarios	Corrective actions
Scenario 1 (S1)	Add 1 nurse
Scenario 2 (S2)	Add 1 doctor
Scenario 3 (S3)	Add 1 nurse + 1 doctor
Scenario 4 (S4)	Add 1 nurse + 1 box
Scenario 5 (S5)	Add 1 doctor + 1 box
Scenario 6 (S6)	Add 1 nurse + 1 doctor + 1 box

assumed that the time required to detect the perturbation (T_D) and the time required to identify and select corrective actions (T_R) are equal to zero ($T_D = T_R = 0$). Figure 6.9 summarizes the results of the corrective actions in the case of scenarios S2 and S5. According to this figure, the time required by the PED to recover its normal state after disturbance (T_{Rec}) is 121 min for scenario S2 and 76 min for scenario S5. We can therefore note that the T_{Rec} is reduced to 3 h and 15 min in scenario S2, and 2 h in the case of the scenario S5. The corresponding strain situation case model (*SSCase* model) for the best scenarios is presented in Table 6.4.

6.7.2   Results in the Case of Reactive Control

In this part we adapted the simulation model used in proactive control by adding control function to monitor strain indicators as explained below. The objective here is to define the corrective actions for the reactive control. The simulated corrective actions tend to improve the strain indicator. We expose results on P_W indicator.

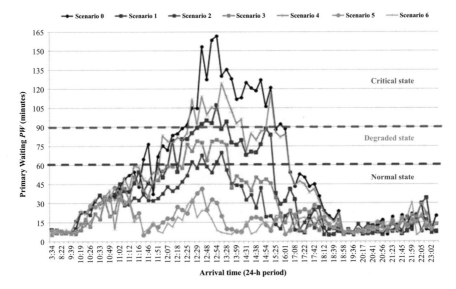

Fig. 6.8 Primary waiting time (P_W) at the PED with corrective actions

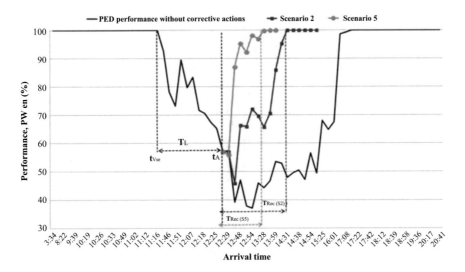

Fig. 6.9 Evolution of PED performance $P_W(t)$

Table 6.4 Proposed corrective actions

C	O	S_I	S_{ED}	V_{TH}	P_{CA}	A_{CA}	T_{REC}
Epidemic period weekend	Minimize or reduce P_W	P_W	Degraded	$P_W \geq 60min$	$S1, S2, S3, S4, S5, S6$	$S2$	121 min

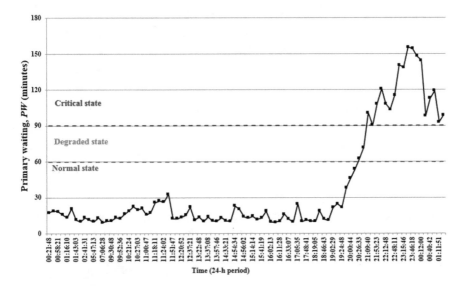

Fig. 6.10 Initial identified ED behavior

The identified state of ED without corrective actions is named *A0* and presented in Fig. 6.10.

The simulation model is used in the short term only to evaluate the impact of defined corrective actions by medical staff. If a new corrective action is identified, medical staff can test it directly on the real system. The simulation can be used in proactive mode to evaluate this new action and its effectiveness to be added to the knowledge base.

The strain indicator P_W is used for monitoring the ED and detecting strain situation:

- If at t_0, the V_{Mo} threshold is exceeded, then the ED is in degraded state.
- The P_W monitoring continues during $\delta(t)$ period. Three cases can be observed:

  - *Case1*: if the indicator continues to increase and exceeds $(V_{Mo}+\delta V_{Mo})$, then a corrective action must be launched.
  - *Case2*: if at $(t_0+\delta t)$ exceed V_{Mo}, then a corrective action must be launched.
  - *Case3*: if at $(t_0+\delta t)$, P_W is lower than V_{Mo} is exceeded, a corrective action must be launched.

- If at t_0 the threshold V_{Cr} is exceeded, a corrective action must be launched.

In this subsection we expose results on corrective actions focused on human and material resources to react to strain situations identified above:

- *Actions on human resources*: add a nurse and/or doctor during a given period (in our case: 2 h).

Table 6.5 Proposed corrective actions

Scenarios	Nurse	Doctor	Box	T_L
A1 to A7	0 or 1	0 or 1	0 or 1	$T_L = 0$ min (Altern. A)
B1 to B7	0 or 1	0 or 1	0 or 1	$T_L = 60$ min (Altern. B)
C1 to C7	0 or 1	0 or 1	0 or 1	$T_L = 90$ min (Altern. C)

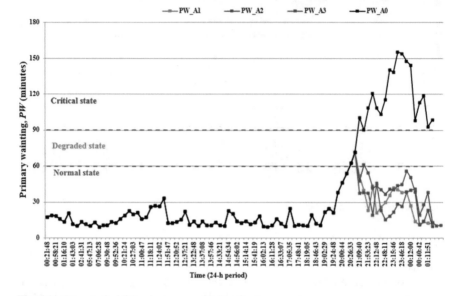

Fig. 6.11 Impact of additional resource with $T_L = 0$ min

- **Actions on material resources**: transform one room in the Short-term Hospitalization Unit (SHU) to a consultation box.

Three alternatives were defined according to the launch time of the corrective action(s) (T_L). Table 6.5 summarizes the different characteristics of these corrective actions. 7 scenarios were defined and assessed for each alternative A (for $T_L = 0$ min), B (for $T_L = 60$ min), and C (for $T_L = 90$ min). 21 scenarios are obtained.

As an illustration, some results are given below for the strain indicator P_W and for each alternative to show the impact of each corrective action and its launch time.

1. **Alternative A : $T_L = 0$** : We present the results of the following scenarios in Fig. 6.11 compared to the identified state $A0$.

   - A1: adding a nurse, the other parameters are equal to 0.
   - A2: adding a doctor, the other parameters are equal to 0.
   - A3: transform one room in the Short-term Hospitalization Unit (SHU) into a consultation box, $T_L = 90$ min, the other parameters are equal to 0.

   As observed in Fig. 6.11, having additional human resources (nurse or doctor) or additional box immediately contributes to the improvement of P_W.

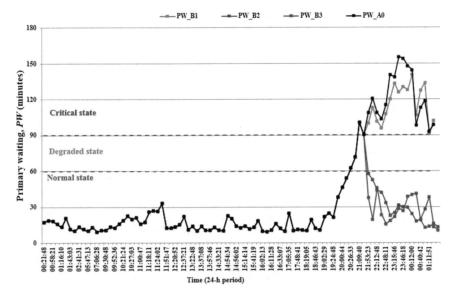

Fig. 6.12 Impact of additional resource with $T_L = 60$ min

2. **Alternative B : T_L = 60 min**: We present the results of the following scenarios in Fig. 6.12 compared to the identified state A0 and in the case of T_L = 60 min:

   - *B1*: adding a nurse, the other parameters are equal to 0.
   - *B2*: adding a doctor, the other parameters are equal to 0.
   - *B3*: transform one room in the Short-term Hospitalization Unit (SHU) into a consultation box, T_L = 60 min, the other parameters are equal to 0.

   As observed in Fig. 6.12, having additional doctor or additional box contributes significantly to the decrease of P_W. Unfortunately, adding a nurse 60 min after the detection of strain situation has no consequence to improve our indicator.

3. **Alternative C : T_L = 90 min**: Fig. 6.13 presents the results of the following scenarios compared to the identified state A0 and in the case of T_L=90 min:

   - C1: adding a nurse, the other parameters are equal to 0.
   - C2: adding a doctor, the other parameters are equal to 0.
   - C3: transform one room in the Short-term Hospitalization Unit (SHU) into a consultation box, T_L = 90 min, the other parameters are equal to 0.

   Transform one room in the Short-term Hospitalization Unit (SHU) into a consultation box" reduces the primary waiting time (PW) of patients in the PED and avoid to the ED to go to critical state. Adding a doctor contributes to improve PW but does not avoid critical state. Adding a nurse after 90 min of strain situation detection has no impact on PW improvement.

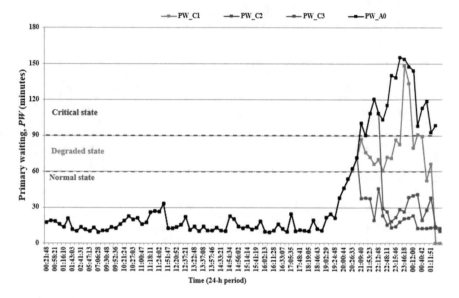

Fig. 6.13 Impact of additional resource with $T_L = 90$ min

6.7.3 Discussion

From the results presented above, we observed the impact of various corrective actions on the behavior of the PED. The launch time of corrective actions plays a key role in some cases, especially in the case of adding nurse(s). This is important to plan nurse's activity and negotiate their shifts and over-time works. We also studied the scenarios combining different types of corrective actions. The results obtained show that certain combinations improve several strain indicators. Some results were expected, but their confirmation and quantification using the proposed models is important for the PED staff.

6.8 Conclusion and Perspectives

The objective of this work is to improve the management of strain situations that may occur in an emergency service. To achieve this goal, we offer a decision support system, ED-CS, to anticipate and manage these situations using simulation and forecasting models. The main objective of this system is twofold: to improve the reception of emergency patients and facilitate the work of staff. The overall results show the interest of those responsible for an ED and also to have a management system providing proactive and reactive management of an ED. This system avoids the occurrence of strain situations and also limits their impact if they do occur.

It also helps to better adapt an organization in terms of human and material resources. However, this study raises several questions about related series. Another important way to improve ED-CS is to continue the analysis and modeling of decision processes and optimize the combination of proactive and reactive control modes. One aspect of this objective is the development of multicriteria algorithms, establishing the priority of a patient according to his or her parameters, based on the degree of severity, pathology, waiting time after his or her arrival at PED, the patient residence time, as well as the stress on people accompanying children. More generally, it seems important to reflect on the approach to be implemented so that the ED is actually a High Reliability Organization (HRO). This problem leads to formalize more precisely the notion of resilience of an ED to consider a methodology for evaluating its resilience indicators. Human factors (including stress for patients and staff) in the care of patients is a fundamental resilience of EDs. It should also be noted that: if we tested a large number of scenarios, it will also be necessary to analyze those that can really be implemented in reality, taking into account the organization of human resources, as well as the regulatory and economic aspects.

Acknowledgments This Chapter is dedicated to our colleague and friend Pr. Christian TAHON. This work is conducted in the HOST project (https://anr.fr/Projet-ANR-11-TECS-0010) and is supported by the ANR (Agence Nationale de la Recherche) of the French Ministry of Research. Special thanks go to the medical and paramedical staff at the PED at CHRU-Lille for their collaboration in this research. Thank you to AIP Nord-Pas-de-Calais in Valenciennes for providing us with ARENA 14.70 (https://www.uphf.fr/aipnpdc/pole-aip-primeca-nord-pas-de-calais).

References

1. C. M. Clancy: Emergency Departments in Crisis: Opportunities for Research. Health Serv. Res, vol. 42, no 1, p. 13–20, 2007.
2. M. Crosse: Hospital Emergency Departments: Crowding Continues to Occur, and Some Patients Wait Longer Than Recommended Time Frames. DIANE Publishing, 2010.
3. D. J. Griffin: Hospitals: What They Are and How They Work. Jones & Bartlett Publishers, 2011.
4. N. R. Hoot and D. Aronsky: Systematic review of emergency department crowding: causes. effects, and solutions , Ann. Emerg. Med, vol. 52, no 2, p. 126–136, 2008.
5. IOM: Hospital–Based Emergency Care: At the Breaking Point. Washington, DC. Washington, DC: The National Academies Press, 2007.
6. A. L. Kellermann: Crisis in the Emergency Department. N. Engl. J. Med, vol. 355, no 13, p. 1300–1303, 2006.
7. L. Rose et al.: Emergency department length of stay for patients requiring mechanical ventilation: a prospective observational study. Scand. J. Trauma Resusc. Emerg. Med, vol. 20, p. 30, 2012.
8. J. R. Twanmoh et G. P. Cunningham: When overcrowding paralyzes an emergency department. Manag. Care Langhorne Pa, vol. 15, no 6, p. 54–59, 2006.
9. IMNA: Institute of Medicine Committee on the Future of Emergency Care in the U.S. Health System. Hospital– Based Emergency Care: At the Breaking Point. The National Academies Press, Washington, DC, 2006.

10. A. Boyle, K. Beniuk, I. Higginson, P. Atkinson: Emergency Department Crowding: Time for Interventions and Policy Evaluations. Emerg. Med. Int, vol. 2012, p. 1–8, 2012.
11. F. Kadri, S. Chaabane, F. Harrou, C. Tahon: Time series modelling and forecasting of emergency department overcrowding. Journal of Medical Systems, p. 38:107, 1–20, 2014.
12. C.-C. Lin, H.-F. Liang, C.-Y. Han, L.-C. Chen, C.-L. Hsieh: Professional resilience among nurses working in an overcrowded emergency department in Taiwan. Int. Emerg. Nurs, vol. 42, p. 44–50, 2019.
13. F. Kadri, S. Chaabane, C. Tahon: A simulation–based decision support system to prevent and predict strain situations in emergency department systems. Simul. Model. Pract. Theory, vol. 42, p. 32–52,2014.
14. L. Aboueljinane, E. Sahin, Z. Jemai, J. Marty: A simulation study to improve the performance of an emergency medical service: Application to the French Val–de–Marne department. Simul. Model. Pract. Theory, vol. 47, p. 46–59, 2014.
15. A. Al-Refaie, R. H. Fouad, M.-H. Li, M. Shurrab: Applying simulation and DEA to improve performance of emergency department in a Jordanian hospital. Simul. Model. Pract. Theory, vol. 41, p. 59–72,2014.
16. H. Beaulieu, J. A. Ferland, B. Gendron, P. Michelon: A mathematical programming approach for scheduling physicians in the emergency room. Health Care Manag. Sci, vol. 3, no 3, p. 193–200, 2000
17. S. L. Doyle, J. Kingsnorth, C. E. Guzzetta, S. A. Jahnke, J. C. McKenna, K. Brown, : Outcomes of implementing rapid triage in the pediatric emergency department. J. Emerg. Nurs. JEN Off. Publ. Emerg. Dep. Nurses Assoc, vol. 38, no 1, p. 30–35, 2012.
18. C. J. Gonzalez, M. Gonzalez, N. M. Ríos :Improving the quality of service in an emergency room using simulation–animation and total quality management. Comput Ind Eng, vol. 33, no 3–4, p. 97–100, 1997.
19. N. Litvak, M. van Rijsbergen, R. J. Boucherie, M. van Houdenhoven: Managing the overflow of intensive care patients. Eur. J. Oper. Res, vol. 185, no 3, p. 998–1010, 2008.
20. J. P. Oddoye, D. F. Jones, M. Tamiz, P. Schmidt: Combining simulation and goal programming for healthcare planning in a medical assessment unit. Eur. J. Oper. Res, vol. 193, no 1, p. 250–261, 2009.
21. J.-Y. Yeh et W.-S. Lin: Using simulation technique and genetic algorithm to improve the quality care of a hospital emergency department. Expert Syst. Appl, vol. 32, no 4, p. 1073–1083, 2007.
22. Z. Zeng, X. Ma, Y. Hu, J. Li, D. Bryant: A simulation study to improve quality of care in the emergency department of a community hospital. J. Emerg. Nurs. JEN Off. Publ. Emerg. Dep. Nurses Assoc, vol. 38, no 4, p. 322–328, 2012.
23. C. Duguay et F. Chetouane: Modeling and Improving Emergency Department Systems using Discrete Event Simulation. SIMULATION, vol. 83, no 4, p. 311–320, 2007.
24. Z. Huang, J. M. Juarez, H. Duan, H. Li: Reprint of "Length of stay prediction for clinical treatment process using temporal similarity". Expert Syst. Appl, vol. 41, no 2, p. 274–283,2014.
25. R. K. Khare, E. S. Powell, G. Reinhardt et M. Lucenti: Adding more beds to the emergency department or reducing admitted patient boarding times: which has a more significant influence on emergency department congestion? Ann. Emerg. Med, vol. 53, no 5, p. 575–585, 2009.
26. M. Laskowski, R. D. McLeod, M. R. Friesen, B. W. Podaima, A. S. Alfa: Models of Emergency Departments for Reducing Patient Waiting Times. PLoS ONE, vol. 4, no 7, p. 1–12, 2009.
27. T. Ruohonen, P. Neittaanmaki, J. Teittinen: Simulation Model for Improving the Operation of the Emergency Department of Special Health Care. in Simulation Conference, 2006. WSC 06. Proceedings of the Winter, 2006, p. 453–458.
28. S. Samaha, W. S. Armel, D. W. Starks: The use of simulation to reduce the length of stay in an emergency department. in Simulation Conference, 2003. Proceedings of the 2003 Winter, 2003, vol. 2, p. 1907–1911.
29. S. Brenner, Z. Zeng, Y. Liu, J. Wang, J. Li, P. K. Howard: Modeling and Analysis of the Emergency Department at University of Kentucky Chandler Hospital Using Simulations. J. Emerg. Nurs, vol. 36, no 4, p. 303–310, 2010.

30. K. A. Dowsland: Nurse scheduling with tabu search and strategic oscillation. Eur. J. Oper. Res, vol. 106, no 2–3, p. 393–407, 1998.
31. A. Komashie et A. Mousavi: Modeling emergency departments using discrete event simulation techniques. in Simulation Conference, 2005 Proceedings of the Winter,2005, p. 2681–2685.
32. E. Marcon, A. Guinet, C. Tahon: Gestion et performance des systèmes hospitaliers. Paris: Hermès science; Lavoisier, 2008.
33. M. D. Rossetti, G. F. Trzcinski, S. A. Syverud: Emergency department simulation and determination of optimal attending physician staffing schedules. In Simulation Conference Proceedings, 1999 Winter,1999, vol. 2, p. 1532–1540.
34. M. Thorwarth, A. Arisha, P. Harper: Simulation Model to Investigate Flexible Workload Management for Healthcare and Servicescape Environment. in Proceedings of the 2009 Winter Simulation Conference, Austin, Texas, USA,2009, p. 1946–1956.
35. T. M. Amaral et A. P. C. Costa: Improving decision-making and management of hospital resources: An application of the PROMETHEE II method in an Emergency Department. Oper. Res. Health Care, vol. 3, no 1, p. 1–6, 2014.
36. E. Cabrera, M. Taboada, M. L. Iglesias, F. Epelde, E. Luque: Simulation Optimization for Healthcare Emergency Departments. Procedia Comput. Sci, vol. 9, p. 1464–1473, 2012.
37. E. Cabrera, M. Taboada, M. L. Iglesias, F. Epelde, et E. Luque: Optimization of Healthcare Emergency Departments by Agent–Based Simulation. Procedia Comput. Sci, vol. 4, p. 1880–1889, 2011.
38. F. Kadri, S. Chaabane, C. Tahon : Reactive Control System to Manage Strain Situations in Emergency Departments. in Proceedings of the 13th International Conference on Informatics in Control, Automation and Robotics, Lisbon, Portugal,2016, p. 576–583.
39. F. Kadri et al.: Modelling and management of the strain situations in hospital systems using un ORCA approach. IEEE IESM, 28–30 October , RABAT – MOROCCO,2013, p. 10,
40. DD. F. Lobach, K. Kawamoto, K. J. Anstrom, M. L. Russell, P. Woods, D. Smith: Development, deployment and usability of a point–of–care decision support system for chronic disease management using the recently–approved HL7 decision support service standard. Stud. Health Technol. Inform, vol. 129, no Pt 2, p. 861–865, 2007.
41. H. Ltifi: Démarche centrée utilisateur pour la conception de SIAD basś sur un processus d'Extraction de Connaissances à partir de Données. Université de Valenciennes, 2011.
42. M. A. Ahmed et T. M. Alkhamis: Simulation optimization for an emergency department healthcare unit in Kuwait. Eur. J. Oper. Res, vol. 198, no 3, p. 936–942, 2009.
43. Y. N. Marmor et al.: Toward simulation–based real–time decision–support systems for emergency departments. In Simulation Conference (WSC), Proceedings of the 2009 Winter, p. 2042—2053, 2009.
44. M. Thorwarth: A Simulation–based Decision Support System to Improve Healthcare Facilities Performance – Elaborated on an Irish Emergency Department. Dublin Institute of Technology, School of Management, 2011.
45. M. Taboada, E. Cabrera, E. Luque, F. Epelde, M. L. Iglesias: A Decision Support System for Hospital Emergency Departments Designed Using Agent–based Techniques. In Proceedings of the Winter Simulation Conference, Berlin, Germany, p. 359:1–359:2, 2012.
46. Z. Y. Zhuang, C. L. Wilkin, A. Ceglowski: A framework for an intelligent decision support system: A case in pathology test ordering. Decis. Support Syst, vol. 55, no 2, p. 476–487,2013.
47. F. Kadri: Contribution à la conception d'un système d'aide á la décision pour la gestion de situations de tension au sein des systèmes hospitaliers. Application à un service d'urgence. Université de Valenciennes et de Hainaut Cambrésis, 2014.
48. J. F. Lebraty et I. Pastorelli-Nège: Biais cognitifs: quel statut dans la prise de décision assitée?. Systèmes Inf. Manag, vol. 9, no 3, p. 87–116, 2004.
49. B. Weckel, Décider face à la complexité et l'urgence. Techniques de l'ingénieur. AG 1 572 2010.: Décider face à la complexité et l'urgence. Techniques de l'ingénieur. AG 1 572, 2010.
50. F. Harrou, F. KADRI, S. Chaabane, C. Tahon, Y. Sun: Improved Principal Component Analysis for Anomaly Detection: Application to an Emergency Department. Comput. Ind. Eng, 2015.

51. I. Hwang, S. Kim, Y. Kim, C. E. Seah:A Survey of Fault Detection, Isolation, and Reconfiguration Methods;, IEEE Trans. Control Syst. Technol, vol. 18, no 3, p. 636–653,2010.
52. S. J. Qin: Survey on data–driven industrial process monitoring and diagnosis. Annu. Rev. Control, vol. 36, no 2, p. 220–234, 2012.
53. V. Venkatasubramanian, R. Rengaswamy, S. N. Kavuri, K. Yin: A review of process fault detection and diagnosis: Part III: Process history based methods , Comput. Chem. Eng, vol. 27, no 3, p. 327–346,2003.
54. X. Dai et Z. Gao: A Data–Driven Perspective of Fault Detection and Diagnosis. IEEE Trans. Ind. Inform., vol.9, no 4, p. 2226–2238, 2013.
55. R. Isermann: Fault–Diagnosis Systems – An Introduction from Fault Detection to Fault Tolerance. ,Springer, 2006.
56. M. Basseville et I. V. Nikiforov: Detection of Abrupt Changes: Theory and Application. Upper Saddle River, NJ, USA: Prentice–Hall, Inc., 1993.
57. A. Mouzakitis: Classification of Fault Diagnosis Methods for Control Systems , Meas. Control, vol. 46, no 10, p. 303–308, 2013.
58. R. Zemouri: Contribution à la surveillance des systèmes de production à l'aide des réseaux de neurones dynamiques: Application à la e–maintenance. Université de Franche–Comté, 2003.
59. F. Harrou, M. N. Nounou, H. N. Nounou, M. Madakyaru: Statistical fault detection using PCA–based GLR hypothesis testing. J. Loss Prev. Process Ind, vol. 26, no 1, p. 129–139, 2013.
60. J. J. Hisnanick: Forecasting the demand for inpatient services for specific chronic conditions. J. Med. Syst, vol. 18, no 1, p. 9–21, 1994.
61. S. S. Jones, A. Thomas, R. S. Evans S. J. Welch: Forecasting daily patient volumes in the emergency department. Acad. Emerg. Med. Off. J. Soc. Acad. Emerg. Med, vol. 15, no 2, p. 159–170, 2008.
62. M. Xu, T. C. Wong, K. S. Chin: Modeling daily patient arrivals at Emergency Department and quantifying the relative importance of contributing variables using artificial neural network. Decis. Support Syst, vol. 54, no 3, p. 1488–1498, 2013.
63. P. D. Babin et A. G. Greenwood: Discretely evaluating complex systems: simulation is a valuable tool for lean Six Sigma. Ind. Eng. Manag, vol. 43, no 2, p. 34–38, 2011.
64. L. Aboueljinane, E. Sahin, Z. Jemai: A review on simulation models applied to emergency medical service operations. Comput. Ind. Eng, vol. 66, no 4, p. 734–750, 2013.
65. A. Belaidi, B. Besombes, E. Marcon: Réorganisation d'un service d'urgences et aide au pilotage des flux de patients?: apport de la modélisation d'entreprise et de la simulation de flux , Logistique Manag, vol. 15, no 1, p. 61–73, 2007.
66. M. Diefenbach et E. Kozan: Hospital emergency department simulation for resource analysis. Industrial Engineering & Management Systems, 2008. [http://eprints.qut.edu.au/30940/].
67. M. Gul et A. F. Guneri: A comprehensive review of emergency department simulation applications for normal and disaster conditions. Comput. Ind. Eng, vol. 83, no 0, p. 327–344, 2015.
68. S. S. Jones et R. S. Evans: An Agent Based Simulation Tool for Scheduling Emergency Department Physicians. AMIA. Annu. Symp. Proc, vol. 2008, p. 338–342, 2008.
69. Y.-H. Kuo, O. Rado, B. Lupia, J. M. Y. Leung, C. A. Graham: Improving the efficiency of a hospital emergency department: a simulation study with indirectly imputed service–time distributions. Flex. Serv. Manuf. J, p. 1–28, 2014.
70. M. Taboada, E. Cabrera, F. Epelde, M. L. Iglesias, E. Luque: Using an Agent–based Simulation for Predicting the Effects of Patients Derivation Policies in Emergency Departments. Procedia Comput. Sci, vol. 18, p. 641–650, 2013.
71. D. Sinreich et Y. Marmor: Emergency department operations: The basis for developing a simulation tool. IIE Trans, vol. 37, no 3, p. 233–245, 2005.

Chapter 7
A Decentralized Approach to the Home Healthcare Problem

Brahim Issaoui, Anis Mjirda, Issam Zidi, and Khaled Ghédira

Abstract In this chapter, we try to improve health care services, particularly the Home Health Care Problem (HHCP). In fact, Home Health Care Service (HHCS) is known as a care mode allowing patients who suffer from complex and evolving diseases to benefit at home from medical and paramedical coordinated care that can be only provided in hospitals. In this chapter, we treat the Caregivers' Tours Problem (CTP) and conflict management sanitary visits to patients' homes. We developed a new distributed three-phase approach, which both optimizes the daily caregivers' tours to minimize the travel costs and maximizes the planned services in order to address potential conflicts. The obtained numerical results, compared with those provided by the other existing approaches, are motivating and encouraging.

7.1 Introduction

For thirty years, the number of beds in private clinics and hospitals has decreased. Meanwhile, the population aging has led to an increase in the number of people suffering from chronic degenerative diseases and given rise to functional disabilities as well as handicaps. Home Health Care Service (HHCS) is essential in our life since it watches over the human being health. In fact, the economic conditions of elderly health care have oriented those persons to the HHCS outside the walls of the hospital. Patients, undergoing treatments for evolutionary chronic diseases or

B. Issaoui (✉) · I. Zidi
Complex Outstanding Systems Modelling Optimization and Supervision (COSMOS), Strategie d' Optimisation et Informatique intelligentE (SOIE), National School of Computer Science (ENSI), Manouba, Tunisia
e-mail: brahim.issaoui@ensi.rnu.tn

A. Mjirda
University of Sfax, Faculty of Economic Science and Management, Sfax, Tunisia

K. Ghédira
Université Centrale de Tunis, Honoris United Universities, Tunisia
e-mail: khaled.ghedira@universitecentrale.tn

© Springer Nature Switzerland AG 2021
M. Masmoudi et al. (eds.), *Operations Research and Simulation in Healthcare*, https://doi.org/10.1007/978-3-030-45223-0_7

palliative care, want a care as close as possible as from their family to get more moral support. Over time, this kind of service has aroused the curiosity of several researchers in the field of industrial engineering and operational research. The Home Health Care Problem (HHCP) can be divided into two classes as proposed in [15]. These classes are organized around two categories of decision-making, including organizational decisions and those related to the implementation of patient care. The former have been divided into three different levels of decision [12] explained below:

– Strategic level: It is necessary to know the needs of the structure in human and material caregivers as well as the skills of the staff who will be recruited.
– Tactical level: It is to manage the skills of human caregivers as well as staff size and assign human caregivers who visit patients in the regions.
– Operational level: It is related to problems arising from the assignments of activities caregivers, staff planning tours, deliveries of medicines, etc.

This chapter deals with conflict caregivers' tours for the HHCP. The distributed proposed approach consists in finding a daily schedule of home health care that minimizes both the caregivers workday durations and the conflict of management sanitary visits to patients' homes. This chapter is organized as follows: In the second section, we present the problem description and aims. Next in the following section, we mention some research works that have contributed to improving the HHCP. We detail the distributed approach in the fourth section. In the fifth section, we analyze and discuss the numerical obtained results and compare them with those provided in the literature. The sixth section is devoted to cite some recommendations to better use the proposed system. Finally, we end up the chapter with conclusions and some prospects.

7.2 Problematic and Aims

As mentioned in the introduction, humanity is currently leaning toward a new mode of care, which is HHC [1–3]. Companies providing HHCS often have problems with care planning and limitations regarding care strategies. We specifically address the problem of conflicting visits when assigning care plans.

We mean by conflictual visit that two (or more) caregivers visiting the same patient at the same time. So imagine the caregivers' problems that affect the entire functioning of the health plan. The centralized approaches do not treat the conflicts' problems between caregivers because they do not address the constraint that says: a patient can have multiple visits by deferent caregivers in the same day. In other words, we are facing an HHCP as it exists patients shared between some caregivers proposed for that day, which is described in Fig. 7.1.

Figure 7.1 illustrates the complexity of determining the caregiver's route for a basic case of seven patients and four caregivers. Each caregiver has a set of patients to visit (e.g., Emmanuel = Johan, David, Keven, Wayne. The four caregivers are

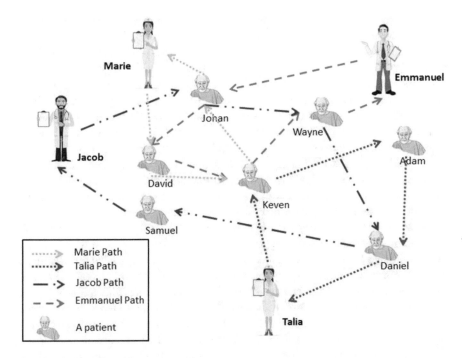

Fig. 7.1 Patients shared between caregivers

linked because some have the same patients. Hence, planning the route for each caregiver implies taking constraints linked to the other caregivers into account. To plan Emmanuel's route, it is necessary to take into account the constraints of the caregivers Jacob and Marie (because they share patient Johan) and of the caregiver Talia (she shares patient Keven). Similarly, if the caregiver Jacob has to care for patient (Daniel) before the caregiver Talia, Talia's route will be linked and constrained by that of Jacob. Our goal in this work is to contribute to enhance the HHCS by proposing a new heuristic to solve the issue of the conflict health visits using a multiagent system, which both optimizes the daily caregivers' tours to minimize the travel costs and maximizes the planned services in order to address potential conflicts.

7.3 Literature Review

We focus here on the recent and similar works dealing with the tours' problem with synchronized or/and shared visits. We begin with the work by Gayraud et al. [4]. They treated the Vehicle Routing Problem (VRP) with synchronization to organize home care system. The authors defined a mathematical model based on the

Multiple Travelling Salesman Problem with Time Windows (m-TSPTW) having specific constraints, such as the timing of activities or patient preference. They also proposed a resolution method using a meta-heuristic. The authors solved the conflict problem with a comparison made on the service history. The assignment of caregivers to patients relies on the preferences of patients, resulting in a significant number of services that cannot be offered despite the interesting patient satisfaction. A heuristic, based on simulated annealing for the VRPTW with synchronized visits, was treated by [5]. This algorithm integrates multiple local search techniques adapted to timing constraints named Simulated Annealing with Iterative Local Search (SA-ILS). Their problem is a similar line to our raised issue, but they did not address the constraint of customer downtime despite its efficiency in terms of run time because they encode their algorithm in C ++ under Linux environment.

Bredström and Rönnqvist projected the problem of caregivers tours to a Synchronized VRPTW [6]. They contributed to the resolution of Synchronized VRPTW problem by defining a mixed integer programming (MIP) solved first with CPLEX, and second by a heuristic. The method solves only small instances, but their heuristic solved the problem effectively by relaxing several hard constraints that are related to time windows and load balancers. Another work dealing with the problem of tours and based on a VRP variant is VRPTW with Temporal Dependencies made by Dohn et al. [7]. The authors have introduced two compact formulations relying on the decomposition of Dantzig–Wolfe and the method of column generation. Their approach solved the problem perfectly for tours to a maximum of 25 customers (patients) who have shared more than 2 visits. The applied time window branching was combined with time window reductions to restore feasibility with respect to temporal dependencies. For computational testing, they have developed a fourth model, which is a hybrid of the relaxed formulation and the time-indexed formulation.

Liu et al. [8] focused on the problem of tours for the HHCS. The discussed issue concerns the supply of medicines and medical devices of the pharmacy home care companies in patients' homes, the administration of special drugs for hospital patients, and collection of biological samples and unused drugs and patients' medical devices. In their research study, two whole joint programming models are offered. Then, they proposed a Genetic Algorithm (GA) and a Tabu Search (TS) Algorithm to resolve their issue. The former is based on a permutation chromosome, a procedure division, and a local search. However, the latter relies on the attributes of patient route assignment, a function of increased cost, reoptimizing road, and attribute levels of aspiration.

Rasmussen et al. [9] have treated the problem of caregivers' tours as a scheduling issue. The problem was modelled as a set of subproblems with side constraints. The model was implemented using column generation in Branch and Price framework. Their approach has improved distances, but not the number of the rendered services, because the objective of minimizing distances has more priority than the management of the shared visits.

Riazi et al. [21] have proposed to apply a distributed Gossip algorithm to solve the tours and the scheduling problems entirely for structures providing HHC. They also applied an extended version called n-gossip. It provided the flexibility for balancing optimality with respect to the calculation speed. The authors have also suggested a relaxation of the underlying solver optimality in the Gossip. This relaxation accelerates the iterations of the Gossip algorithm and allows obtaining the good quality solutions in shorter time. Their goal is focused on the effectiveness of their algorithm and robust calculation. They tested all Gossip algorithm variants to choose the best one. Unfortunately, the tests were performed on small instances (20, 25, and 30 jobs and 3, 4, and 5 caregivers).

Trautsamwieser and Hirsch [11] have introduced a two-step solution for real-life scenarios using meta-heuristic based on the Variable Neighborhood Search (VNS). The first step generates initial schedules for the caretakers by assigning the jobs without violating the constraints. The second step applies VNS to refine the schedules obtained by the initial solution and to find the combination that minimizes the total cost. The main advantage of this approach is that the formulation criteria make the solution more applicable in a real-life scenario, e.g., ensuring that breaks respect work regulations. Furthermore, by using soft time window constraints for the majority of jobs and minimizing the violation of time window with regard to time window violations, some slack is allowed to come up with a feasible solution. The main weakness of this approach is that there is no way to find a global optimum, or even a feasible solution using standard solver software for real-life situations (i.e., larger problems) without applying heuristics. The staff tours problem was also discussed by Thomson in [12], where it was considered as vehicle routing problem with time window in which the author has taken into account the shared visits (two caregivers for the same visit). The objective of Thomson' s proposal is to reduce the total travel time and increase the number of visits carried out by the permanent staff. His approach is based on mathematical modelling. In order to minimize the travel time and the waiting time for patients, Eveborn et al. [13] have introduced a tool on the basis of optimization methods and heuristics. The problem was generally formulated. It consists in assigning to each employee a schedule among all possible schedules, so that all visits will be carried out by minimizing the total cost. Each tour has a particular cost depending on time and journey duration, work schedules, individual preferences, and unfulfilled constraints. In terms of resolution, two successive couplings were made. Initially, each visit and each professional were, respectively, associated with a tour. Merging the two elements (visit and professional) was done using heuristic rules and a local search. The solutions got in minutes have reduced the working time by 7% (significantly reducing the duration of daily meetings for the realization of manual tours) and the duration of trips by 20%.

De Angelis [14] dealt with the caregivers allocation problem for home care case, more specifically the case of AIDS patients. He has proposed a mathematical modelling by linear programming for the caregivers allocation between patients needing home care (medical, social assistance, or support for patients). De Angelis distinguished two types of problems: a local problem of caregivers allocation and an overall problem concerning the budget to be allocated to the public health service for home medical care.

However, authors in [15] developed a multiobjective mixed integer programming mathematical model focusing on: the increase of patient satisfaction, the number of health interventions, and the reduction in the care intervention costs. To solve their problem, the authors used a 3-step solution. A new Mixed Integer Linear Programming (MILP) model has been proposed by [10] to make a planning for a home health care problem. The model is optimizing routes and rosters for the health care staffs, while problem-specific constraints are satisfied. This model integrates an original concept related to the human behavior (e.g., patient behavior).

You can also see these three studies [17, 18], and [22] in which authors give two literature reviews about the HHC.

To conclude, it is obvious that the tours problem of the caregivers and the conflict visits can be seen as a Vehicle Routing Problem with Discrete Split Delivery and Time Windows (VRPSDTW) [19].

In the next section, we will discuss our approach.

7.4   Proposed Approach

First, we will describe the proposed approach by a Mixed Integer Linear Programming (MILP) model. Second, we will explain and describe the distributed resolution as a second part.

7.4.1   Mathematical Model

We use two very well-known models found in the literature [10, 16] in order to define our model, which is summarized as follows:

Parameters

K: set of caregiver
R: set of requests
P: set of patients
$[ar, br]$: time windows to serve the request $r \in R$
τ_r: service time to serve the request $r \in R$
l_r: localization of the request $r \in R$, $L = \{l_r : r \in R\}$
t_{ij}: travel time associated with the arc $(i,j) \in E$
d_0: main depot (the company)
d_k: home depot of the caregiver $k \in K$, $D = d_k : k \in K$
$G = (V, E)$: a directed graph, where $V = D + L + 0$ is the node set

$$S_r^\tau = \begin{cases} 1 & \text{if the caregiver k has the competence to serve the request r} \\ 0 & \text{otherwise} \end{cases}$$

Decision Variables

$$x_{ij}^k = \begin{cases} 1 \text{ if the caregiver k traverses the arc (i,j)} \\ 0 \text{ otherwise} \end{cases}$$

$$y_r^k = \begin{cases} 1 \text{ if the caregiver k is assigned to the request r} \\ 0 \text{ otherwise} \end{cases}$$

U_r^k =starting time of service by caregiver k at request r (or depot home).

Mathematical Formulation

$$Minimize \sum_{i=1} \sum_{ij \in E} t_{ij} x_{ij}^k \tag{7.1}$$

Subject to

$$\sum_{k \in K} y_r^k = 1 \qquad\qquad \forall r \in R \tag{7.2}$$

$$\sum_{j \in L + d_0} x_{d_k,j}^k = 1 \qquad\qquad \forall k \in K \tag{7.3}$$

$$\sum_{j \in L} x_{j,d_k}^k = 1 \qquad\qquad \forall k \in K \tag{7.4}$$

$$\sum_{j \in L + d_0} x_{j,l_r}^k = y_r^k \qquad\qquad \forall k \in K \forall r \in R \tag{7.5}$$

$$\sum_{j \in L + d_0} x_{l_r,j}^k - \sum_{j \in L + d_0} x_{j,l_r}^k = 0 \qquad\qquad \forall k \in K \forall r \in R \tag{7.6}$$

$$y_r^k \leq S_r^k \qquad\qquad \forall k \in K \forall r \in R \tag{7.7}$$

$$a_k \leq u_{dk}^k \leq b_k \qquad\qquad \forall k \in K \tag{7.8}$$

$$a_r \leq u_{lr}^k \leq b_k \qquad\qquad \forall k \in K \forall r \in R \tag{7.9}$$

$$u_i^k + \tau_i + t_{ij} - M(1 - x_{ij}^k) \leq u_j^k \qquad\qquad \forall k \in K \forall i, j \in V \tag{7.10}$$

$$u_r^k + M(1 - y_r^k) \geq a_r \qquad\qquad \forall k \in K \forall r \in R \tag{7.11}$$

$$u_r^k + \tau_r - M(1 - y_r^k) \leq b_r \qquad\qquad \forall k \in K \forall r \in R \qquad (7.12)$$

$$x_{ij}^k \in \{0, 1\} \qquad\qquad \forall ij \in V \forall k \in K \qquad (7.13)$$

$$y_r^k \in \{0, 1\} \qquad\qquad \forall r \in R \forall k \in K \qquad (7.14)$$

The objective function (1) minimizes the total duration (distance) of all routes performed by the caregiver. The constraint (2) ensures that each request is fulfilled exactly once. The constraints (3) and (4) mean that each caregiver starts from home depot and ends at home depot, respectively. The constraint (5) means that if a request r is assigned to a caregiver k, then the localization of this request must be visited by this caregiver, while the constraint (6) corresponds to the flow constraints. The constraint (7) means that a caregiver cannot be assigned to a request if their skills do not correspond to the required skills to satisfy the request. Constraints (8–12) satisfy the corresponding time windows of caregiver and patients.

7.4.2 Decentralized Approach

The HHCP is distributed by nature, because caregivers and patients are distant. We propose a distributed approach based on a multiagent system that contributes to solving the problem of conflicting visits. Our approach is founded on a negotiation policy between the caregivers and the patients and on a conflict management rule.

7.4.2.1 Conflicts Management Rule

In the following example in Fig. 7.2, we will explain the principle of the rule: Considering two caregivers Cg1 and Cg2 who have two lists of medical tasks with different sizes and intervention schedules. In Fig. 7.2, we note that there is an overlap

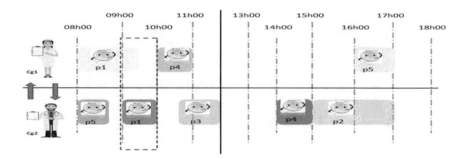

Fig. 7.2 Time overlap

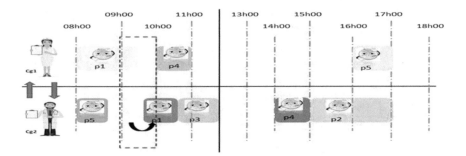

Fig. 7.3 Solution for time overlap

of tasks, which corresponds the patient P1. The resolution is in Fig. 7.3, in which for example the caregiver Cg2 executes the Conflicts Management Rule by performing a different time to serve the patient 1. Although, this rule may be insufficient due to the fact that only one caregiver should execute it. Thus, executing this rule risks by ending up in a new conflict situation. This invites us to answer the question: who should apply this delay rule?

To answer to this question, we define between caregivers some negotiation policies that are discussed in the following section.

Although, this rule may be insufficient due to the fact that only one caregiver should execute it. Thus, executing this rule risks by ending up in a new conflict situation. This invites us to answer the question: who should apply this delay rule? We have answered this question by developing intercaregivers' negotiation policies that are detailed in the next section.

7.4.2.2 Negotiation Policies

In this chapter, we have developed two different and independent policies (POL1 and POL2). We called the first policy POL1: Number of remaining patients to visit; the second one is called POL2: caregiver who has the most flexible calendar.

7.4.2.3 POL1 «Number of Remaining Patients to Visit (NRP)»

In Fig. 7.4, we detail the NRP policy for the conflict situation, for example, between two caregivers. The caregiver Cg1 will move her visit to the patient P1 presenting the conflict situation, because the number of her visits to perform after the conflict is the smallest, then the potential risk of creating new conflictual situation will be minimum.

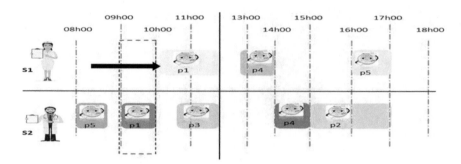

Fig. 7.4 Solution provided by POL1:NRP

7.4.2.4 POL2 «Overall Equipment Effectiveness (OEE)»

For this rule, the caregiver will move his visit to whoever has the most flexible
calendar; in other words, who is the one with the lowest Overall Equipment
Effectiveness (OEE). In fact, the OEE is calculated as the ratio between the «total
duration of care» and the «total travel time» OEE = Total care duration / Total travel
time. See [20]. Here in the Fig. 7.5 we have

– Caregiver1 has:

  Total care duration == 03h30
  Travel time ==3h (e.g., p4 and p5 are too far apart)
  OEE = 1.16

– Caregiver2 has:

  Total care duration == 6h
  Travel time ==02h30 (e.g., p2, p3, p4 are close)
  OEE = 2.4

Hence, the change is effected by the caregiver1 because 1.16<2.4.

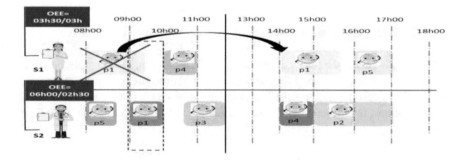

Fig. 7.5 Solution provided by POL2:OEE

We integrated each negotiation policy into a Multiagent System (MAS). We will describe the agents' types in the next section before discussing how they communicate between each other.

7.4.3   Agents' Types

The MAS cited in Fig. 7.6 was composed by a "Home Health Care Center Agent" (HHCC Agent) and a set of "Subsystems": caregivers' agents and patients' agents.

7.4.3.1   Agent Home Health Care Center

Agent Home Health Care Center represents the HHC facility. He/she provides to the patients: the lists of caregivers, health actions, and the required durations for care.

7.4.3.2   Caregiver Agent

The caregiver agent represents the caregiver who makes visits to patients' homes. And he/she performs two types of communications: (i) with another caregiver, in case of conflict problem, to negotiate and find a conflict management decision; (ii) with patients to collect their availabilities and/or to offer hours of passage at their homes, etc.

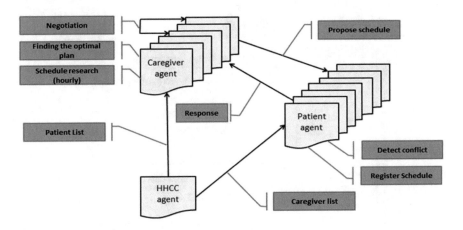

Fig. 7.6 Overview of the system

7.4.3.3   Patient Agent

After receiving his daily list of caregivers, this agent remains in waiting state till having proposals of passing' hours of all caregivers. He checks these times: if there is not a conflicting schedule, he/she answers his/her caregivers that everything is well by saving those time periods. Otherwise, he/she transfers the conflict problem to the concerned caregivers in order to resolve the trouble.

7.4.3.4   Agents' Communications

Figure 7.7 summarizes the whole scenario that could be happening.
   The messages exchanged between the agents are explained as follows:

Message 1:   the HHC Center agent informs the different caregiver agents about the proposal patient agents and the proposal duration of care.

Message 2:   the HHC Center agent informs the different patient agents about the proposal caregiver agents and the proposal duration of care.

Message 3:   caregiver agents look for the availabilities of patient agents because each patient agent has a different availability from another one, which complicates resolution of the problem and makes more and more conflict schedule planning.

Message 4:   Patient agents provide information corresponding to their availability to the caregiver agents.

Message 5:   caregiver agents solve the problem (the first resolution) based on the first shared data.

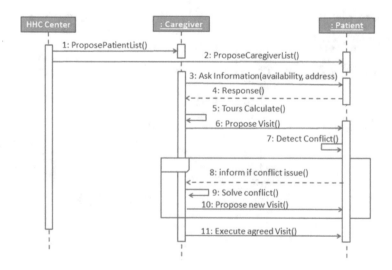

Fig. 7.7 Sequence diagram: interaction between members of the system

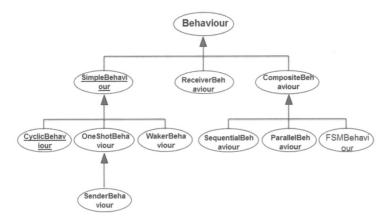

Fig. 7.8 Different possible behaviors

Message 6: after the first resolution made by caregivers basing on the local information collected from the local patient, each caregiver agent informs his/her patient agents about the proposal visits.

Message 7: after receiving the proposal schedule, patient agents are invited to detect the conflict situation.

Message 8: Patient agent informs the concerning caregiver agents about the conflict problem.

Message 9: here the caregiver agents negotiate between themselves in order to solve one more time the conflict situation by applying a negotiation policy (one from the two explained in Sect. 7.4.2.2).

Message 10: caregiver agents inform the patient agent about the new schedule.

Message 11: caregiver agents inform the patient agent that the problem was successfully resolved and tell him/her about the new schedule or tell him/her that it is impossible to visit.

Figure 7.8 shows different behaviors existing in the literature.

We have a bidirectional communication between caregiver agents and patient agents, and we need them rest a live as possible as that why we choose a cyclic behavior for the different agents in our MAS.

7.5 Computational Results

In this part of this chapter, we present the computational results obtained after performing several tests to show the efficiency of our approach and two referent approaches applied on different benchmarks offered by Aiane et al. in [10] and Dohn et al. in [16].

The proposed approaches were coded in Java on Toshiba with Intel®Core™Duo CPU T6500 with 2.1 GHz processor core 2 and 4 GB of RAM.

7.5.1 First Comparison

We represent the comparison between (POL2, and POL1) and the (MILP) approach proposed by Aiane et al. in [10] regarding three criteria α, β, and γ such as:

- α: number of detected conflict;
- β: % resolved conflict;
- γ: travel time in hour.

These three approaches were tested on three kinds of benchmarks in Tables 7.1, 7.2, and 7.3: (A): 9 small instances (2 to 4 caregivers, 8 to 12 patients, and 14 to 20 requests for care), (B): 5 medium instances (5 to 10 caregivers, 15 to 30 patients, and 45 to 87 requests for care), (C): 6 large instances (10 to 15 caregivers, 30 to 50 patients, and 110 to 200 requests for care).

Table 7.1 Small instances (A)

	OEE			NRP			MILP		
	α	β	γ	α	β	γ	α	β	γ
S1	0	100	409	0	100	409	0	100	409
S3	0	100	585	0	100	585	0	100	585
S5	0	100	715	0	100	715	0	100	715
S21	0	100	815	0	100	815	0	100	815
S23	0	100	925	0	100	925	0	100	925
S25	0	100	906	0	100	906	0	100	906
S31	0	100	908	0	100	908	0	100	908
S33	0	100	1085	0	100	1085	0	100	1085
S35	0	100	1095	0	100	1095	0	100	1095
Average	0	100	827	0	100	827	0	100	827

Table 7.2 Medium instances (B)

	OEE			NRP			MILP		
	α	β	γ	α	β	γ	α	β	γ
M211	98	91	523	95	100	602	Not converged	Not converged	Not converged
M212	86	84	761	91	99	743	Not converged	Not converged	Not converged
M341	102	85	785	115	100	712	Not converged	Not converged	Not converged
M342	116	91	815	118	100	836	Not converged	Not converged	Not converged
M485	101	94	905	105	97	882	Not converged	Not converged	Not converged
Average	100.6	89	757.8	104.8	99.2	755	Not converged	Not converged	Not converged

Table 7.3 Large instances (C)

	OEE			NRP			MILP		
	α	β	γ	α	β	γ	α	β	γ
B5318	148	79	772	169	94	758	Not converged	Not converged	Not converged
B8034	137	74	874	149	92	891	Not converged	Not converged	Not converged
B9237	156	76	785	174	89	741	Not converged	Not converged	Not converged
B100A	175	81	921	170	96	978	Not converged	Not converged	Not converged
B100B	169	71	714	197	93	784	Not converged	Not converged	Not converged
B100C	163	86	956	205	97	911	Not converged	Not converged	Not converged
Average	158	77	837	177	93	843	Not converged	Not converged	Not converged

In fact, β represents the major criterion in order to choose the best approach because our primary goal is to resolve the conflict between the caregivers. In our study, we note that the approaches that adopt NRP and OEE gave the same performances as [10] on (A). And this is explained by the small size of (A)'s instances.

Regarding instances (B) and (C), we note that the approach based on NRP solves the conflict problem with better resolution percentages than the other approaches. But by examining α (number of conflicts detected), the approach based on NRP is more conflictual than the other approaches, because it makes a lot of changes to solve the conflict problem. For the last comparison criterion γ (travel time in hours), the approach based on OEE gives a better result than the approach based on NRP (Table 7.4).

7.5.2 Second Comparison

A second comparison with the proposed approach by Dohn et al. in [16] was done. The criteria taken into consideration this time are as follows:

- ϕ: Uncovered visits;
- ψ: Total travel time;
- λ: Run time.

Tests are performed on four benchmarks provided by [16] and represented in Table 7.5: hh : 150 visits, 15 Caregivers; II1: 99 visits, 8 Caregivers; II2: 60 visits, 7 Caregivers; II3 : 61 visits, 6 Caregivers. To simplify the analyses of the above tests, we computed the numerical averages in the Table 7.4.

We note that the approach based on NRP covers the maximum number of visits compared to the other approaches. Then the approach based on OEE comes second. Also, we note that OEE solves the problem with fewer distances than NRP, and this is so logical, because OEE covers less health visits compared to NRP. But, for all

Table 7.4 Numerical averages of (*A*), (*B*), and (*C*)

	OEE			NRP			MILP		
	α	β	γ	α	β	γ	α	β	γ
A	0	100	827	0	100	827	0	100	827
B	100.6	89	757.8	104.8	99.2	755	Not converged	Not converged	Not converged
C	158	77.8	837	177.3	93.5	843.8	Not converged	Not converged	Not converged
Average	86.2	88.94	807.26	94.04	97.56	808.61	Not converged	Not converged	Not converged

Table 7.5 Second comparison

	Clustering			OEE			NRP		
	ϕ	ψ	λ	ϕ	ψ	λ	ϕ	ψ	λ
hh	5	448	244	6	431	36	4	462	22
II1	6	280	69	3	274	19	1	291	10
II2	0	141	30	1	116	17	0	135	8
II3	1	128	39	0	92	12	0	108	9
Average	3	249.2	95.5	2.5	228.2	21	1.2	249	12.2

test instances hh, II1, II2, and II3, we note that the approach based on NRP covers the maximum number of visits compared to the other approaches.

7.5.3 Discussions

As a synthesis, the approach based on NRP is the most effective on the three comparison criteria and on the data set compared to the other approaches developed and proposed in the literature. Tests performed on 5.1 confirm that NRP is a very conflictual policy and in 5.2 invites us to more optimize the run-time solving.

7.5.4 Example of Execution

In Table 7.6, we present the final result obtained by the execution of P1:NRP on 14 caregivers and 25 patients and 87 home health interventions. «a» and «b» represent, respectively, the start and the end of health care time intervention. We present all necessary information regarding the schedule for caregivers and patients especially the order of tour for the caregiver (i.e., the schedule of his/her visits).

7.6 Recommendations for Reuse

In this part of the chapter, we give some recommendations in order to better use the proposed system. First, the caregiver must be mobile, hence the need to have a permit and a car if no professional vehicle is made available.

Second, a good physical constitution is required to be able to properly handle patients. The caregiver must also be psychologically sound because he or she is often alone in the face of human distress (aggressiveness, death, etc.).

Table 7.6 Example of final result calculated by the system

	P1	P2	P3	P4	P5	P6	P7	P8	P9	P10	P11	P12	P13	P14	P15	P16	P17	P18	P19	P20	P21	P22	P23	P24	P25
a		750			900			1185		1050						840			1125		990				
b		795			945			1200		1080						855			1140		1005				
S1*		1			3			7		5						2			6		4				
a			615						840					690	915									750	
b			645						870					705	960									795	
S2*			1						4					2	5									3	
a								1030		940			765			700			850						
b								1050		985			805			720			895						
S3*								5		4			2			1			3						
a		885	945	810						1155						750	1095			1010					
b		900	965	840						1200						765	1110			1050					
S4*		3	4	2						7						1	6			5					
a	1180								985				1055				895				1115		815		
b	1200								1010				1070				940				1135		850		
S5*	6								3				4				2				5		1		
a		1125					960					810										1035			885
b		1140					990					840										1080			915
S6*		5					3					1										4			2
a			880				760					820		970						1105		1180			
b			925				775					835		1000						1135		1200			
S7*			3				1					2		4						6		7			
a	980				1040			755					1185				1100	830			890				

(continued)

Table 7.6 (continued)

	995	1055	785	1200	1140	845	935
b							
S8*	4	5	1	7	6	2	3
a	975	1170	755	1035	890	1110	825
b	990	1200	780	1065	930	1125	845
S9*		4			3	5	2
a	690	605			765	885	750
b	720	645			780	930	770
S10*		1	4		3	6	
a		585	855		735	945	
b		615	900		750	960	
S11*	2	1		5	4	3	
a		750	1035	690	825	975	
b		780	1050	705	870	990	
S12*		2	6	1	5	3	
a		900	1035	960	815	609	
b		915	1050	990	855	705	
S13*		4		6	3	1	2
a		590		780	720	870	930
b		615		825	735	885	960
S14*		1	2	4	3	5	6

Finally, the caregiver must have a minimum knowledge of some computer tools, especially tablets or smart bracelets, to do their job more efficiently and especially to use our proposed system.

7.7 Conclusions and Future Works

In this chapter we proposed a distributed HHCS while minimizing the number of conflicts that may arise when defining schedules of visits to the patients' homes. We have developed two decentralized approaches. Each one is based on a policy of independent negotiation and then the application of the task shift. After studying the various calculated numerical results, we concluded that the policy based on NPR gives better results in terms of resolving detected conflicts than the other developed policy and approaches proposed in the literature.

As a future work, we opt to refine the proposed approach to reduce work's overtime while keeping the same quality of conflict resolution. Also, we aim to enhance our approach by taking into account the stochastic scenarios, such as stochastic events that could inhibit the HHCP normal operations (e.g., traffic jam, the lack of caregiver, and so on).

References

1. Guinet, A : Organisation des soins à domicile en Europe et en Amérique du nord. 10th International Conference on Modelling, Optimization and Simulation. Nancy, France, November 05–07, 2014.
2. Bashir, B., Chabrol, M., Caux, C.: Literature review in home care. 9th International Conference of Modelling, Optimization and Simulation. Bordeaux, France, June 06–08, 2012.
3. Woodward, CA., Abelson, J., Tedford, S., Hutchison, B.: What is important to continuity in home care? Perspectives of key Stakeholders. Social Science & Medicine. vol 58, 177–192 (2004).
4. Gayraud, F., Deroussi, L., Grangeon, N., Norre, S.: Problème de tournées de vhicule avec synchronisation pour l'organisation de soins à domicile. 10th International Conference on Modelling Optimization and Simulation. Nancy, France, November 05–07, 2014.
5. Afifi, S., Dang, D., Moukrim, A.: Une heuristique pour un problème de tourneés de véhicules avec fenêtres de temps et des visites synchronisées [Paper in French]. 16ème congrès annuel de la Société française de Recherche Opérationnelle et d' Aide à la Décision (ROADEF), 2015.
6. Bredström, D., Rönnqvist, M.: Combined vehicle routing and scheduling with temporal precedence and synchronization Constraints. European Journal of Operational Research. vol 191 (1), 19–31, (2008).
7. Dohn, A., Rasmussen, MS., Larsen, J.: The vehicle routing problem with time windows and temporal dependencies. Networks , vol 58 (4), 273–289, (2011).
8. Liu, R., Xie, X., Augusto, V., Rodriguez, C.: Heuristic algorithms for a vehicle routing problem with simultaneous delivery and pickup and time windows in home health care. European Journal of Operational Research, 230 (3), 475–486, (2013).

9. Rasmussen, MS., Justesen, T., Dohn, A., Larsen, J.:The home care crew scheduling problem : Preference-based visit clustering and temporal dependencies. European Journal of Operational Research, **219** (3), 598–610, (2012).
10. Aiane, D., El-Amraoui, A., Mesghouni, K.: A new optimization approach for a home health care problem. 6th IEEE International Conference on Industrial Engineering and Systems Management. Seville, Spain, October, 2015.
11. Trautsamwieser, A., Hirsch, P.: A branch-price-and-cut approach for solving the medium-term home health care planning problem. Networks, **64** (3), 143–159, (2014).
12. Thomson K 2006. Optimization on home care. PhD thesis, Informatics and Mathematical Modelling, Technical University of Denmark, DTU.
13. Eveborn, P., Flisberg, P., Ronnqvist, M.: LAPS CARE an operational system for staff planning of home care. European Journal of Operational Research, **171**, 962–976, (2006).
14. De Angelis, V.: Planning Home Assistance for AIDS Patients in the City of Rome. Interfaces **28**, 75–83, (1998).
15. Issaoui, B., Zidi, I., Marcon, E., Ghedira, K.: New multi-objective approach for the home care service problem based on scheduling algorithms and variables neighbourhood descent. Electronic Notes in Discrete Mathematics. **47**, 181–188, (2015), doi:10.1016/j.endm.2014.11.024
16. Dohn A, Rasmussen MS, and Larsen J. 2009b. The vehicle routing problem with time windows and temporal dependencies. PhD thesis, Informatics and Mathematical Modelling, Technical University of Denmark.
17. Issaoui, B, Zidi, I., Marcon, E., Laforest, F., Ghedira, K.: Literature review: home health care. 15th International Conference on Intelligent Systems Design and Applications. Marrakesh, Morocco, December, 485–492, 2015.
18. Bashir, B., Caux, C.: Literature review in Home Care. 9th International Conference of Modelling and Simulation. Bordeaux, France, June, 2012.
19. Archetti, C., Speranza, MG.: Vehicle Routing Problem with Discrete Split Delivery and Time Windows. International Transactions in Operational Research. **19**, 3–22, (2012).
20. Issaoui, B., Zidi, I., Ghedira, K.: A New Metaheuristic for the Home Health Care Problem: Caregivers Tours and Conflict Visits. IEEE International Conference on Systems, Man, and Cybernetics (SMC). Budapest, Hungary, October, 2016.
21. Riazi, S., Chehrazi, P., Wigström, O., Bengtsson, K., Lennartson, B.: A gossip algorithm for home healthcare scheduling and routing problems. The International Federation of Automatic Control. Cape Town, South Africa, **19**, 10754–10759, (2014).
22. Christian F., Patrick H.: Home health care routing and scheduling: A review. Computers and Operations Research, **77**, 2017, 86–95, ISSN 0305-0548, https://doi.org/10.1016/j.cor.2016.07.019.

Chapter 8
Wounded Transportation and Assignment to Hospital During Crisis

Mohamad Khorbatly, Hamdi Dkhil, Hassan Alabboud, and Adnan Yassine

Abstract Disasters have catastrophic effects on human lives that require a speedy evacuation reaction in order to minimize their severe consequences. In a crisis management supply chain, the main objective of our study is to minimize the total transporting time of wounded in order to minimize the human loss. Evacuation is a tool to save human lives, but it is not always a practicable decision to protect the lives of residents. Therefore, a prompt disaster response, i.e., emergency evacuation planning, involves the planning of the required resources assignment and timely vehicle scheduling, which are compulsory to organize successful secured operations that could, if well organized, save many injured humans and lessen the human suffering.

To achieve this goal, we treat the Integrated Problem of Ambulance Scheduling and Resources Assignment (IPASRA) in the case of a sudden disaster. The required resources are the ambulances and the hospitals. However, the hospitals' serving capacities might be considered or not according to the extent of disaster and particularly to the wounded bodies' total number. We formulate the (IPASRA) as a linear model; furthermore, a novel hybrid algorithm based on Tabu Search (TS) and Greedy Randomized Adaptive Search Procedure (GRASP) is offered to tackle this complex problem.

M. Khorbatly (✉)
ISSAE-CNAM University, Tripoli, Lebanon
e-mail: mohamad.khorbatly@isae.edu.lb

H. Dkhil
LMAH, Normandie Univ, Unihavre, Le Havre, France
e-mail: hamdi.dkhil@univ-lehavre.fr

H. Alabboud
Faculty of Economics and Business Administration - Section 3, Lebanese University, Tripoli, Lebanon

A. Yassine
Institut Supérieur d'Etudes Logistiques, Normandie Univ, Unihavre, Le Havre, France
e-mail: adnan.yassine@univ-lehavre.fr

© Springer Nature Switzerland AG 2021
M. Masmoudi et al. (eds.), *Operations Research and Simulation in Healthcare*, https://doi.org/10.1007/978-3-030-45223-0_8

8.1   Introduction

Violent natural disasters have been a fact of human life since the beginning of the species. Due to the sudden occurrence of disasters, the reaction and the ability to make practical decisions must be quick and efficient as much as possible in order to minimize the human loss by providing quick help and effective rescue to the injured people. Over the past few years, many natural and triggered disasters had occurred and led to hundreds and thousands of wounded and killed people. China was hit by severe floods between January and July, killing at least 144 people—56 alone over several days in early July 2017 [1], and other natural disasters killed an additional 70 people in the first half of 2017; all disasters have displaced about 1 million people and destroyed 31,000 homes. A major 7.1 magnitude earthquake shook Mexico City on September 19, 2017 [2], killing at least 225 people. The earthquake struck on the 32nd anniversary of a devastating 1985 quake that killed thousands in Mexico and came less than two weeks after another massive earthquake killed at least 96 people and left 2.5 million in need of aid. Mexico rescinded its pledge for Hurricane Harvey relief after that 8.2 magnitude quake, which destroyed or damaged tens of thousands of homes. A 7.5-magnitude earthquake struck the Southern Highlands province of Papua New Guinea in February 2018 [3], triggering a major aftershock and some landslides. Close to half a million people were affected by the disaster. In September 2018, flooding in Nigeria displaced more than half a million people and destroyed more than 13,000 homes. The floods struck one-third of Nigeria's 36 states, affecting nearly 2 million people [4]. Torrential rain and landslides impacted large areas of Japan in July 2018 [5], killing more than 200 people. In September 2018, flooding in Nigeria displaced more than half a million people and destroyed more than 13,000 homes. There is also another kind of disasters, which are made by people like battles as happened in Iraq, 2003 [9], when US troops started the invasion, in coalition with the UK and other nations. By the August 31, 2010, when the last US combat troops left, 4,421 had been killed. Almost 32,000 had been wounded in action. Similarly, an enormous loss was reported in Syria since the beginning of the conflict in 2011 [6], around 11.5% of the country's population have been killed or injured since the crisis erupted in March 2011; the report estimates that the number of wounded is put at 1.9 million. Life expectancy has dropped from 70 in 2010 to 55.4 in 2015. While in the Libyan armed conflict 2011 [7] as per Mohamed A. Daw et al. [13], a total of 21,490 (0.5%) persons were killed and 19,700 (0.47%) were injured. Hence, terrorist attacks hit multiple cities all over the world in different ways, such as car bomb, individual attacks, etc., but all of them have the same consequences and common results revealed by a huge number of killed and wounded citizens. All these attacks are considered as human disasters and treated as high-level crises. For example: On the November 13, 2015 [8], six attacks in 33 min hit the capital of Paris, France, from France stadium to the capital center; they caused a death of 129 citizens and hundreds of wounded spreads within this area, most are seriously injured. In this chapter, we propose a mathematical tool for solving the Integrated Problem of Ambulance Scheduling and Resource Assignment (IPASRA) in order to transport wounded as fast as possible to the available medical treatment centers. We present a mathematical modelling succeeded by a Branch and Cut under CPLEX solver and an efficient resolution that outperform meta-heuristic methods and grounded on a hybrid algorithm based on Tabu Search and GRASP procedure.

8.2 Literature Review

In this section we present few definitions of supply chain management and other works dedicated to evacuation only.

Supply Chain Management (SCM) is an essential element to operational efficiency. SCM can be applied to customer satisfaction and company success, as well as within societal settings, including medical missions. Many researchers define it because of the vital role that (SCM) plays within organizations, and hence employers seek employees with an abundance of (SCM) skills and knowledge. This has been defined by many researchers. As per Christopher [10], it is a network of grouped companies that operate in harmony and participate in different processes and activities for creating products and services to end-consumer.

Lummus in 1998 [11] defined that the supply chain is a network of entities in which the material flux passes through these entities, including providers, transporters, assembling sites, distribution centers, retailers, and clients. M. Chen et al. [14] described a linear programming approach to deal with multipriority group evacuation in sudden onset disasters. They assumed that they have in advance the population total number, which cannot be modified later. The authors' goal was to minimize the total travel time. Kai Huang et al. [12]. have developed an integrated multiobjective optimization model for optimizing three objectives: minimizing human loss, optimizing delay cost, and fairness; to achieve this, they combined resources allocation with emergency distribution in disaster response operations. In their study they aim to deliver urgent resources to in-site wounded in adequate quantity as fast as possible within a time barrier, and they do not consider picking up wounded to hospitals or medical treatment centers. Other researchers like Wei Yi et al. [15] were interested in studying the impact of a disaster and presented a meta-heuristic approach grounded on the ant colony optimization to solve the logistics problem of relief activities, which includes wounded evacuation to medical centers and dispatching commodities to distribution centers in crisis zones. Their strategy depends on two phases of decision-making. At first, they build stochastic vehicle paths, and then second, they solve the assignment problem between different types of vehicle flows and commodities. The authors suggest creating temporary medical facilities in different places inside the affected areas in addition to the existing hospitals, and also they plan to provide necessary goods to all existing facilities and transport all wounded toward them. Some of the authors like Felix Wex et al. [16] focused on emergency response in crisis management, and they plan to minimize the total time spent in evacuating wounded outside disaster areas using different rescue units but do not have a plan to transport them to medical units. Mori et al. [17] have planned an evacuation model using multiple types of helicopters to maximize the number of wounded transported to secure areas, and they took the time limit for survival as a hard constraint to be respected; but in their strategy, there was no plan to evacuate all victims to medical treatment centers. Other authors like Byung Hoon Oh et al. [18] study the wounded evacuation problem after a natural disaster, and they were interested in maximizing the number of evacuees by using helicopters with different capabilities. The main inconvenience of using helicopters is that they need a vast location to land that may not be always available and sometime inapplicable. The GRASP-TS technique was first studied by Laguna et al. [19] in 1991. Then it has been useful in many fields, such as location problem with Juan A. Díaz et al. [22] in 2017, Lozano & García-Martínez [20], 2010, and Blum [21], 2010.

8.3   Problem Formulation

The problem is divided into two integrated subproblems; the first subproblem in concerned in treating the scheduling problem, and the second one focuses on resources' assignment problem. The situation is defined as follows: we have many dispersed wounded with different injuries, several ambulances, and several hospitals, and we intend to transport z wounded to m hospital sites with k ambulance. Each hospital h_i treats (theoretically) a finite number p of wounded, and the index i in use is a counter to specify the hospital number. We have to schedule the ambulance–wounded allocation problem depending on wounded serving priority and, at the same time, we plan to optimize the ambulance–hospital assignment process.

The following rules are defined for this revision.

```
The estimated wounded number is known in advance.
We use only ambulances.
We will assume that the total number of ambulances to be
  used for wounded transportation purpose is less than the
  estimated number of wounded.
We will limit the ambulance capacity to one wounded
  at a time.
```
We will limit a finite capacity c_i for a hospital h_i.
```
We will consider that the sum of all hospital's capacities
  is greater than the estimated wounded number.
All wounded are distant from each other.
We will not consider the wasted time spent by the operation
  of picking up and dropping off wounded from their location
  to hospital.
Each hospital is capable to treat all kind of injuries.
```

8.3.1   Mathematical Formulation

We present below the data used in our modelling:

1. H is the set of hospitals indexes and $|H|$ the number of hospitals.
2. A is the set of ambulances indexes and $|A|$ the number of ambulances.
3. H_{fic} and P_{fic} are, respectively, the set of fictive wounded and the set of fictive hospitals. Each vehicle a is associated to a pair (p_a,h_a), where p_a is a fictive wounded and h_a a fictive hospital and where the distance between p_a and h_a is nil. These elements are used to ensure the starting of each vehicle missions. The distance between h_a (or p_a) and each hospital h represents the distance between the depot of ambulance a and the hospital h. In others terms, p_a and h_a represent the depot of ambulance a.
4. w_h : Theoretical capacity of hospital h; it is the maximum number of wounded whom the hospital h can serve at the beginning of the operation.
5. $T_{p,h}$: The total routing time spent in transporting the injured p to hospital h. This matrix is not symmetric in real cases.
6. $T_{h,p}$: The total routing time spent in moving from the hospital h to the wounded p. We can consider here that h is the last visited hospital and p the next wounded to secure.
7. G: A sufficiently large number.

8.3.2   Equations

In the following we introduce the variables of our model:

C_{max} : The date of operation's ending called the makespan. This value is the objective to minimize in our modelling.

t_p : The time when the ambulance transporting wounded p arrives at hospital.

$X^a_{p,h,f}$: Binary variable defined as follows:

$X^a_{p,h,f} \begin{cases} = 1 \text{ if ambulance } a \text{ transfers wounded } f \text{ directly before wounded } p \text{ to hospital } h \\ = 0 \text{ otherwise} \end{cases}$

$$\sum_{p \in P} X^a_{p_a,h_a,p} = 1 \tag{8.1}$$

$$\sum_{a \in A} \sum_{p \in P} \sum_{f \in P} X^a_{p,h,f} = 1 \quad \forall f \neq p \tag{8.2}$$

$$\sum_{a \in A} \sum_{h \in H} \sum_{f \in P} X^a_{p,h,f} = 1 \quad \forall p \in P, \quad \forall f \neq p \tag{8.3}$$

$$\sum_{a \in A} \sum_{h \in H} \sum_{f \in P} X^a_{f,h,p} = 1 \quad \forall p \in P, \quad \forall f \neq p \tag{8.4}$$

$$\sum_{h \in H} \sum_{f \in P} X^a_{p,h,f} = \sum_{l \in H} \sum_{g \, in \, P} X^a_{g,l,p} \quad \forall p \in P, \quad \forall a \in A, \quad \forall f \neq p, \quad g \neq p \tag{8.5}$$

$$t_f >= T_{p,h} + T_{h,f} + t_p + G * (X^a_{p,h,f} - 1)$$

$$\forall a \in A, \quad \forall p \in P, \quad \forall f \in P/P_{fic}, \quad \forall f \neq p \tag{8.6}$$

$$C_{max} >= t_p, \quad \forall p \in P/P_{fic} \tag{8.7}$$

$$X^a_{p,h,f} \in \{0, 1\}, \quad t_p \in R^+, \quad \forall a \in A, \quad \forall p, f \in P, \quad h \in H \tag{8.8}$$

$$R_f >= 1 + R_p + |P| * (X^a_{p,h,f} - 1)$$

$$\forall a \in A, \quad \forall p \in P, \quad \forall f \in P/P_{fic}, \quad \forall f \neq p \tag{8.9}$$

$$C_{max} >= \sum_{p \in P} \sum_{h \in H} \sum_{f \in P/P_{fic}} X^a_{p,h,f}(T_{p,h} + T_{h,f}), \quad \forall a \in A \tag{8.10}$$

$$\sum_{a \in A} \sum_{p \in P} \sum_{f \in P} X^a_{p,h,f} <= w_h, \quad \forall f \neq p, \quad \forall h \in H/H_{fic} \tag{8.11}$$

Constraint (1) ensures the starting of each ambulance mission by associating it to a fictive couple of injured and hospital. Note that these two fictive elements are situated in the same location as the associated ambulance at the beginning of its mission. The distance between the fictive injured and the fictive hospital is nil, and the distance between the fictive hospital and each injured is equal to the distance between the initial location of the ambulance and the injured, which is initially known. Constraint (2) ensures that each fictive hospital is to be visited only once. Constraint (3) ensures that he or she has only one direct successor, and constraint (4) ensures that he or she has only one predecessor. In other terms, constraints (3) and (4) ensure that each injured should be transported once. With constraint (5) we ensure that the successor and the predecessor of each injured are transported by the same ambulance. Constraint (6) ensures the temporal succession by considering the binary variable $X^a_{p,h,f}$ and the traveling times $T_{p,h}$ and $T_{h,f}$, with this constraint we evaluate the ending date of each securing mission denoted by t_p. Constraint (7) is used to evaluate the makespan, which is the time spent by the last mission ending. The makespan in denoted by C_{Max}. Constraint (8) represents decision variable. The evaluation of the makespan by constraints (6) and (7) represents a serious complication at simulation stage because of G, which is a sufficiently large number. To improve the modelling, we replace constraints (6) and (7), respectively, by constraints (9) and (10). We replace also t_p by R_p, which represents a number associated to wounded p and used to avoid subcycles.

Constraint (9) is imposed to avoid subcycles, and with constraint (10) the makespan is reduced by the total traveling time of each ambulance. By adopting constraints (9) and (10) instead of constraints (6) and (7), we improved the efficiency of the modelling in terms of simulation results. Note that the use of G in constraint (9) does not affect the quality of simulation results as in the case of constraint (6). Note that we integrated the ambulance scheduling and hospital affectation in the same variable $X^a_{p,h,f}$ to avoid the use of a big number in constraint (10). Considering the presented constraints, the capacity of hospitals is not taken into account. According to crisis extent, the problem can be considered with or without capacity constraint. In fact, for a major disaster, when the number of injured is larger than the total capacity of hospitals, the capacity constraint should be ignored. However, in other situation, when the number of injured is less than the total capacity of all hospitals, it is more realistic to consider the capacity constraint in the modelling in order to propose an adequate solution. Capacity constraint is modelled by the following inequality. Our linear model is solved using CPLEX solver 12.6.

8.3.3 Figure

As explained before, we deal with the Integrated Problem of Ambulance Scheduling and Resource Assignment where the resource assignment concerns the allocation of hospitals and ambulances to wounded. The situation is described as follows: we have a finite number of dispersed wounded, a fleet of ambulances located in several ambulatory centers, and several hospitals in distant areas, and we intend to transport the set of wounded to the hospital sites with the fleet of ambulances (Fig. 8.1).

8.4 Problem Solution

In this section we solve our linear model using CPLEX solver 12.6 that counts on branch-and-cut method, which consists in the combination between the cutting plane method and the branch-and-bound algorithm. We find that:

$$\text{Branch \& Cut} = \text{Branch \& Bound} + \text{Cutting Planes}.$$

8.4.1 Proof of NP-HARDNESS

Consider a particular instance of IPASRA where the number of hospitals |H| is equal to the number of wounded |P| and where the distance between each injured and the

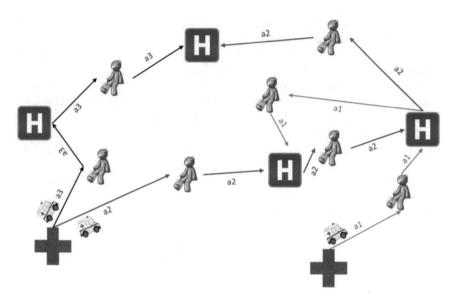

Fig. 8.1 Problem description

Table 8.1 Iteration number

Family	K_{Max}	I_{Max}	Z_{Max}
F1	1000	100	300
F2	10,000	200	1000
F3	1,000,000	300	2000

nearest hospital is negligible compared to the distances between hospitals. For that instance, our problem is equivalent to the Vehicle Routing Problem (VRP) where the vehicles are the ambulances and the cities are the hospitals. If the VRP is known to be NP-Hard, then IPASRA is NP-Hard (Sect. 8.2).

8.4.2 Resolution Method

This section presents the obtained results for the instances of the IPASRA. The algorithm is coded using java, and the computer used for simulation has a CPU Core I7 2.4 GHZ, 16 GB of RAM, 2GB dedicated VGA RAM, and 500 GB SSD hard disk.

In Table 8.1 we present generated instances for simulation tests, and for each one we give the number of wounded, the number of ambulances, the number of hospitals, and finally the capacity of each hospital. Note that for every instance two other data are defined, the distances between hospitals and each wounded and the distances between ambulances and wounded at the beginning of the secure operation. Instances are classified into three families F1, F2, and F3. F1 contains instances with a number of injured equal to 10 or 15 wounded, F2 contains instances

Table 8.2 Problem characteristics

Family	Instance	|P|	|H|	|A|
F1	Inst1	10	2	1
F1	Inst2	10	2	2
F1	Inst3	10	3	2
F1	Inst4	10	3	3
F1	Inst5	15	2	2
F1	Inst6	15	3	2
F1	Inst7	15	3	3
F2	Inst8	20	2	2
F2	Inst9	20	4	2
F2	Inst10	20	3	3
F2	Inst11	20	4	3
F2	Inst12	20	4	4
F2	Inst13	30	5	4
F2	Inst14	50	8	4
F3	Inst15	70	5	4
F3	Inst16	70	8	4
F3	Inst17	70	9	4
F3	Inst18	100	5	5
F3	Inst19	150	10	5
F3	Inst20	200	10	5
F3	Inst21	200	20	5

of 20, 30, or 50 injured, and F3 is the family containing instances of more than 50 wounded.

In Table 8.2, we present the numerical results for capacitated IPASRA. The simulation tests of non-capacitated IPASRA are presented in Table 8.3.

8.4.2.1 Greedy Randomized Adaptive Search Procedure (GRASP)

GRASP is a fast-randomized heuristic used to solve combinatorial optimization problems. It was first proposed by Feo and Resende in 1995 [5], and it consists of two phases: the first one is a repetitive randomized construction and the second one is a local search phase in which the constructed solution is subject to many improvements.

We discuss now the adaptation of GRASP algorithm to solve our problem. As shown in next diagram, we start the algorithm by initializing randomly the initial solution. Then we apply an iterative randomized procedure to construct a feasible solution. At each iteration of the randomized procedure, we treat the subsequent injured to secure and we select randomly, with probability one of the nearest ambulances and one of the nearest hospitals having sufficient capacity. Then we update the distances between the set of injured and the selected ambulance at this iteration. We update also the capacity of the selected hospital. When we finish the

Table 8.3 CPLEX results, capacitated IPASRA

Family	Instance	CPLEX value (sec)	Hybrid value	CPLEX CPU	Hybrid CPU	CPLEX GAP	Hybrid GAP
F1	Inst1	11503	11503	<1	<1	0%	0%
F1	Inst2	2782	2728	2.8	48	0%	0%
F1	Inst3	3776	3776	0.81	6	0%	0%
F1	Inst4	2943	2943	7.64	317	0%	0%
F1	Inst5	2328	2328	172.29	234	0%	0%
F1	Inst6	8719	8721	123.07	48	0%	0.02%
F1	Inst7	1801	1818	355.7	313	0%	0.94%
F2	Inst8	3497	3499	23	24	0%	0.06%
F2	Inst9	8594	8753	3600	164	0.02%	1.85%
F2	Inst10	1762	1777	3600	54	0.51%	0.85%
F2	Inst11	2384	2413	3600	163	0.34%	1.22%
F2	Inst12	6150	6113	3600	94	3.84%	−0.6%
F2	Inst13	3572	3541	3600	111	5.49%	0.87%
F2	Inst14	6590	5779	3600	242	21.88%	−12.31%
F3	Inst15	8433	8866	10800	61	6.11%	5.13%
F3	Inst16	7490	7761	10800	188	9.73%	3.62%
F3	Inst17	7926	7770	10800	200	12.62%	−1.97%
F3	Inst18	N.S	15552	10800	40	–	
F3	Inst19	N.S	23550	10800	965	–	
F3	Inst20	N.S-O.M	20197	9673	866	–	
F3	Inst21	N.S-O.M	7874	<1	820	–	

N.S (No Solution), O.M (Out Of Memory)

treatment of all wounded, we check if the constructed solution is better than all of the previous constructed solutions, if it was the case then we save it as the best solution, and then we reinitialize the parameters of the randomized procedure and we restart it again. The number of randomized constructions named K_{Max} is initially predefined; when it is reached, the procedure is stopped and the best solution found is returned.

Note here that we use the normal uniform law in the random distribution.

8.4.2.2 Tabu Search (TS)

In 1989 and in 1990 Glover [1, 2] was the first who developed a Tabu Search algorithm, which has been effectively powerful and useful for solving combinatorial optimization problems. This technique relies on two phases: a. adaptive memory and b. responsive exploration; the first one records ancient search information depending on four important references: occurrence, recently visited, quality, and influence. The second one is capable to make tactical selections to achieve effectiveness. The main role of using an adaptive memory is to record the previsited good candidates. This record is called tabu list (forbidden list), and its main purpose is to restrict choosing redundant candidates (marked as Tabu active) for some time; depending on the user prefixed Tabu list size, when the list size exceeds the user predefined limit, the first record is removed (aspiration method). The main importance of this adaptive memory is to avoid entering in a local optimum and to enhance the search by using intensification or diversification strategies, where the intensification strategy allows the algorithm to exploit more the good found candidates, while the diversification strategy leads to explore unvisited candidates inside the solution space. We present in the next paragraph the adaptation of Tabu Search approach to our problem.

We provide a detailed explanation for neighborhood construction after this general description. At first, let us define a negative iteration as an iteration that does not improve the best solution found. After that, we use the number of successive negative iterations as a diversification criterion, an intensification criterion, and a stopping criterion. Before launching the Tabu Search procedure, an initial solution is generated and all of the required parameters are initialized such as: the tabu list, the current solution, the best solution, and the number of successive negative iterations. Next, we start the iterations of our algorithm. At each iteration, the neighborhood of the current solution is constructed. The size of this neighborhood depends on the number of successive negative iterations at the current iteration. The current solution is updated by the best non-tabu neighbor; if this neighbor is better than the best explored solution, then the best solution will be updated and the neighbor is injected to the tabu list. If the maximal size of tabu list is reached, the oldest element is removed. When the number of successive negative iterations is equal to a fixed integer, which we denote by I_{Max}, the Tabu Search is stopped and the best solution is returned.

The neighborhood of each constructed solution is divided into three subneighborhoods where each one is constructed according to a particular vision. For the first subneighborhood construction, the method is done by selecting a vehicle and replacing it with another one. The wounded is being inserted in the other vehicle associated cycle. For the second subneighborhood construction, we select a vehicle and we switch only the associated wounded with another wounded; in other words, we switch service order between its associated wounded. For the construction of the third subneighborhood, we select a vehicle and we replace only the associated hospital with another available one and update their capacities.

8.4.2.3 Hybrid Algorithm

Hybrid algorithms are popular and well known for their efficiency compared to classic meta-heuristics. By using hybridization, we can avoid falling in some problems like the stagnation under local optima, and also we can efficiently reduce the required processing time spent in calculating an optimal solution or an acceptable solution. Many hybridization strategies are developed in the literature, but we can identify two-man classes of hybridization. The first one is the hybridization by parallelization or by threading, and it is a strategy based on the exploitation of concurrence and cooperation benefits when many meta-heuristic processes are launched in parallel. The second approach is the sequential hybridization and is based on the alternate between two meta-heuristics or more and on a cooperation technique like the communication among some parameters. The hybrid algorithm that we propose is a sequential hybridization of GRASP algorithm and TS procedure. This approach is based on restarting the Tabu Search from different initial solutions generated by GRASP algorithm for a specific iteration number denoted by Z_{Max}.

Note here that the values of I_{Max}, K_{Max}, and Z_{Max} are subject to modifications depending on the instances size. We present in the next Table 8.1, the values of these iterations according to instances size.

In Table 8.2, we present instances generated for simulation tests, and for each one we give the number of wounded, the number of ambulances, the number of hospitals, and the capacity of each hospital. Note that for every instance two other data are defined, the distances between hospitals and injured and the distances between ambulances and injured at the beginning of the secure operation. Instances are classified into three families F1, F2, and F3. F1 contains instances with a number of injured equal to 10 or 15 wounded, F2 contains instances of 20, 30, or 50 injured, and F3 is the family containing instances of more than 50 wounded.

The deviation between the CPLEX solution and the solution given by hybrid algorithm is calculated by:

$$Gap = ((V.H.A - V.C)/V.C),$$

V.H.A: Value given by Hybrid Algorithm
V.C: Value given by CPLEX

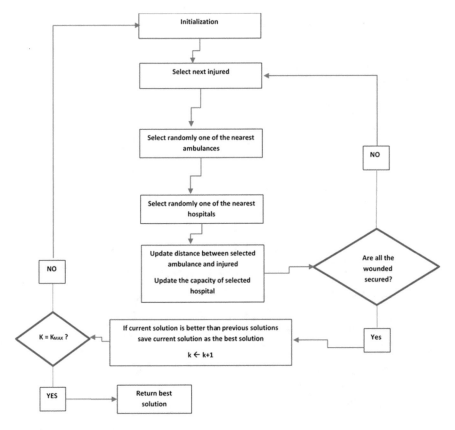

Fig. 8.2 GRASP procedure

With the following diagrams in Figs. 8.2 and 8.3, we present, respectively, our GRASP procedure and hybrid algorithm with the integrated Tabu Search procedure.

Consider now the capacitated IPASRA—Table 8.3. All the solutions given by CPLEX for F1 family are optimal, and the processing time is less than 6 min; in the meantime, the hybrid algorithm provides 5 optimal solutions in less than 3 min 17 s, and for instances Inst6 and Inst7 the gap less than 1%. With F2 family, from a total of 7 instances, only one instance (Inst8) was solved optimally by CPLEX in 23 s, for all other instances CPLEX did not provide optimal solutions after a prefixed CPU processing time of 3600 s, the gap is between 0.34% and 21.88%, and the deviation between CPLEX solutions and solutions given by the hybrid algorithm is less than 2%; all these solutions are given in approximately 4 min. Note here that with instances Inst12 and Inst14 the deviation between these solutions and the solutions given by the hybrid algorithm is, respectively, −0.6% and −12.31%. For Family F3, the CPU processing time is limited to 10800 s, for instances Inst15, Inst16, and Inst17 the CPLEX GAP is, respectively, equal to 6.11%, 9.73%, and 12.62%, and

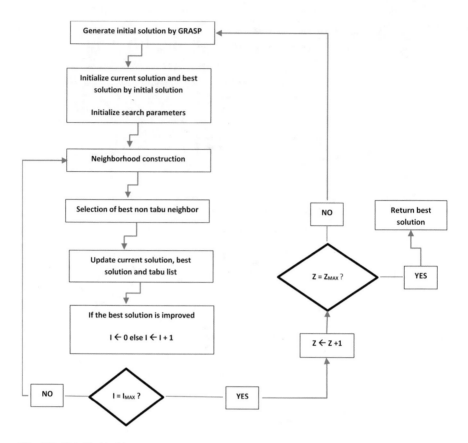

Fig. 8.3 Hybrid algorithm

the deviation between CPLEX solutions and the solutions of hybrid algorithm is, respectively, 5.13%, 3.62%, and −1.97%; these solutions are calculated in less than 3 min 30 s. For instances Inst18 and Inst19, a "No Solution" message was given by CPLEX after 10,800 s of CPU processing time, and for the remaining instances Inst20 and Inst21 a "No-Solution Out Of Memory" message was returned, due to computer hardware limitations, in, respectively, 9673 and less than 1 s.

Note that deviation between CPLEX solutions and hybrid algorithm solutions is negative when the solution given by hybrid algorithm is better than the one given by CPLEX. Note also that solutions given by CPLEX are not optimal for some instances because of memory limit or processing time limitation.

Consider the non-capacitated IPASRA given in Table 8.4. Starting with instances of family F1, all instances were optimal; Inst1, Inst2, and Inst3 are solved in less than 1 sec, Inst4 was solved in exactly 3 s, Inst5 and Inst7 were solved in less than 3 min, and finally Inst6 was solved in less than 17 min. With the hybrid algorithm, from a total of 7 instances, 6 were optimal in less than 3 min, and for Inst6, a

Table 8.4 CPLEX results, non-capacitated IPASRA

Family	Instance	CPLEX value (sec)	Hybrid value	CPLEX CPU	Hybrid CPU	CPLEX GAP	Hybrid GAP
F1	Inst1	11,275	11,275	<1	<1	0%	0%
F1	Inst2	2259	2259	<1	<1	0%	0%
F1	Inst3	3595	3595	<1	<1	0%	0%
F1	Inst4	1878	1878	3	<1	0%	0%
F1	Inst5	2325	2325	69	5	0%	0%
F1	Inst6	8602	8614	1018	38	0%	0.14%
F1	Inst7	1799	1799	170	170	0%	0%
F2	Inst8	2398	2398	3600	9	0.54%	0%
F2	Inst9	7971	7971	3600	164	0.84%	0%
F2	Inst10	852	852	66	54	0%	0%
F2	Inst11	1213	1213	299	163	0%	0%
F2	Inst12	4077	4099	3600	94	1.03%	−0.34%
F2	Inst13	732	732	45	111	0%	0%
F2	Inst14	3416	3288	3600	242	5.50%	−3.75%
F3	Inst15	1732	1832	10800	314	0.06%	2.86%
F3	Inst16	2198	2210	10800	113	1.91%	0.55%
F3	Inst17	9465	3709	10800	121	64.68%	−60.81%
F3	Inst18	N.S	9564	10800	317	–	
F3	Inst19	N.S	17249	10800	305	–	
F3	Inst20	N.S	2629	10800	414	–	
F3	Inst21	N.S-O.M	7874	<1	820	–	

N.S (No Solution), O.M (Out Of Memory)

deviation of 0.14% was given by hybrid algorithm with a processing time equal to 38 s. For instances of F2 family, Inst8, Inst9, Inst12, and Inst14 did not attain optimality after a fixed processing time of 1 h with a gap equal to 0.54%, 0.84%, 1.03%, and 5.5%, respectively. Continuing with F2 family, Inst10, Inst11, and Inst13 were solved optimally by CPLEX in less than 5 min. The deviation between these solutions and solutions given by hybrid algorithm is between −0.34% and −3.75%. Moving to F3 family, for instances Inst15, Inst16, and Inst17 the CPLEX gap is between 0.06% and 64.68%, and these solutions were attained after 10,800 s of CPU consuming time. For instances Inst18, Inst19, and Inst20 there is no solution given by CPLEX after 10,800 s, and finally for Inst21 no solution is returned directly due to "out of memory"; these solutions were attained after 10800 s of CPU consuming time. The deviations between these solutions and the solutions produced by the hybrid algorithm are equal to 2.86%, 0.55%, and −60.81%, respectively.

We should note here that when the deviation between CPLEX solutions and hybrid solutions is negative, it means that the produced solution by hybrid algorithm is better than the ones produced by CPLEX, which are not optimal due to processing time limitation or due to memory limitation.

8.5 Conclusion

The aim of this study is to provide the best ambulance–wounded–hospital scheduling and assignment process, which determines which ambulance, coming from an ambulatory center, should transport a specific wounded to a specific hospital than departing from this hospital to the next wounded in order to transport him or her to another specific hospital, which could be the same one, and so on until all wounded are transported.

At the beginning, we focused on the development of a complete binary linear model that minimizes the total time spent by routing ambulances picking up wounded from their locations to distant hospitals. Then, in a subsequent step, we applied the B&C method using CPLEX solver under JAVA. We applied an effective hybrid meta-heuristic algorithm based on the Tabu Search and GRASP. Two scenarios were treated, and for both of them, the offered model has proved the capability of evacuating wounded from their positions to hospitals within reasonable time.

After discussing the obtained results for both scenarios, and depending on the situation, we can notice the huge importance of choosing between these two scenarios especially with medium instances, the choice should be based on the on-site provided information, especially wounded number and hospitals' theoretical capacities, before taking the final decision. Our model provides in both scenarios optimal solutions for instances with number of wounded up to 50, number of ambulance up to 8, and number of hospitals up to 4.

At the end, the results of the proposed hybrid algorithm are compared with those obtained with the exact method of CPLEX solver. The comparison of the execution time illustrates the efficiency of our proposed model; its results show clearly that it is capable of generating optimal solutions within seconds. The CPLEX resolution, based on our modelling, also proves its efficiency but with more precious time wasted. However, the efficiency of the exact method decreases as the number of instances grows; it is only effective with a medium number of instances.

References

1. F. Glover, Tabu Search Part I. ORSA Journal on Computing, Volume 1, pp. 190–206, 1989, https://reliefweb.int/report/china/natural-disasters-kill-204-china-h1
2. F. Glover, Tabu Search Part II. ORSA Journal on Computing, Volume 2, pp. 4–32, 1990, https://www.usnews.com/news/best-countries/slideshows/10-of-the-deadliest-natural-disasters-of-2017?slide=8
3. https://reliefweb.int/disaster/eq-2018-000020-png
4. http://floodlist.com/africa/nigeria-floods-september-2018-update
5. T. Feo, M. Resende, Greedy randomized adaptive search procedures. Journal of Global Optimization, Volume 6, pp. 109–133, 1995, https://edition.cnn.com/2018/07/10/asia/japan-floods-intl/index.html
6. https://www.theguardian.com/world/2016/feb/11/report-on-syria-conflict-finds-115-of-population-killed-or-injured
7. https://www.sciencedirect.com/science/article/pii/S2211419X15000348
8. http://www.lefigaro.fr/actualite-france/2015/11/14/01016-20151114ARTFIG00252-six-attaques-en-33-minutes-chronologie-d-un-massacre.php
9. http://www.bbc.com/news/world-middle-east-11107739
10. M. Christopher, Logistics and Supply Chain Management, 1st edition, London, Pitmans, 1992.
11. R. R. Lummus, R. J. Vokurka, K. L. Alber, Strategic supply chain planning, Production and Inventory Management Journal, Volume 39 (3), pp. 49–58, 1998.
12. K. Huang et al. K. Huang et al.; Modeling multiple humanitarian objectives in emergency response to large-scale disasters, Transportation Research Part E., Volume 75, pp. 1–17, 2005.
13. Mohamed A. Daw, Abdallah El-Bouzedi, Aghnaya A. Dau, Libyan armed conflict 2011: Mortality, injury and population displacement, African Journal of Emergency Medicine, Volume 5, Issue 3, Pages 101–107, 2015.
14. M. Chen, L. Chen, R. Miller-Hooks, Traffic signal timing for urban evacuation. Journal of Urban planning and development, Volume 133(1), pp. 30–42, 2007.
15. Wei Yi, Linet Ozdamar, A dynamic logistics coordination model for evacuation and support in disaster response activities, European Journal of Operational Research, Volume 179, pp. 1177–1193, 2007.
16. Wex, F., Schryen, G., Feuerriegel, S., Neumann, D. Emergency response in natural disaster management: Allocation and scheduling of rescue units European, Journal of Operational Research, Volume 255, Issue 3, pp. 697–708, 2014.
17. Mori, A.A., Kobayashi, K., Shindo, M. Particle Swarm Optimization / Greedy-Search Algorithm for Helicopter Mission Assignment in Disaster Relief, Journal of Aerospace Information Systems, Volume 12, Number 10, pp. 646–660, 2015.
18. Byung Hoon Oh, Kwangyeon Kim, Han-Lim Choi, Inseok Hwang; Integer Linear Program Approach for Evacuation of Disaster Victims with Different Urgency Levels, IFAC (International Federation of Automatic Control), Volume 50, number 1, pp. 15018–15023, 2017.

19. Laguna, M., & González-Velarde, J. L; A search heuristic for just-in– time scheduling in parallel machines. Journal of Intelligent Manufacturing, Volume 2, Issue 4, pp. 253–260, 1991.
20. Lozano, M., & García-Martínez C.; Hybrid metaheuristics with evolutionary algorithms specializing in intensification and diversification: Overview and progress report. Computers & Operations Research, Volume 37, Issue 3, pp. 481–497, 2010.
21. Blum, C.; Hybrid metaheuristics. Computers & Operations Research, Volume 37, Issue 3, pp.430–431, 2010.
22. Juan A. Díaz, Dolores E. Luna, José-Fernando Camacho-Vallejo, Martha-Selene Casas-Ramírez. GRASP and hybrid GRASP-Tabu heuristics to solve a maximal covering location problem with customer preference ordering, Expert Systems with Applications, Volume 82, pp. 67–76, 2017.

Chapter 9
Carbon Footprints in Emergency Departments: A Simulation-Optimization Analysis

Masoumeh Vali, Khodakaram Salimifard, and Thierry Chaussalet

Abstract It is globally accepted to act against global warming through the reduction of carbon dioxide. Carbon footprint is historically defined as the total emissions caused by an individual, event, organization, or product, expressed as carbon dioxide equivalent. Healthcare system consumes large amount of energy in order to provide health services to patients who have to pass a series of treatment processes at each care unit. These treatments require different medical equipment that consume electrical power, and the more electrical power consumption is, the more greenhouse gases specifically CO_2 emissions are. The discrete-event simulation has been applied to develop the model of the treatment process and the estimation of carbon dioxide in the treatment process. By the knowledge that the simulation is not an optimization method in itself, the OptQuest optimization method has been applied to reduce greenhouse gases and carbon footprint in the *patients'* flow in the emergency department by considering leveling off the waiting time and length of stay as constraints to leveling up *patient's* satisfaction. The numerical results provided by simulation and OptQuest show the efficiency of OptQuest as a technique for patient flow optimization.

9.1 Introduction

It is globally accepted to act against global warming through the reduction of carbon dioxide [1]. There is a wide range of greenhouse gases (GHGs), including nitrogen oxide, methane, F-gases, and also carbon dioxide (CO_2) proportionally responsible for GHG, for example, the share of nitro oxide is about 8%, while methane is blamed for 16% of GHG. The *earth's* climate is changing, owing to greenhouse

M. Vali (✉) · K. Salimifard
Department of Industrial Management, Persian Gulf University, Bushehr, Iran
e-mail: m.vali@mehr.pgu.ac.ir; salimifard@pgu.ac.ir

T. Chaussalet
Computer Science and Engineering Department, Westminster University, London, UK
e-mail: chausst@westminster.ac.uk

© Springer Nature Switzerland AG 2021
M. Masmoudi et al. (eds.), *Operations Research and Simulation in Healthcare*, https://doi.org/10.1007/978-3-030-45223-0_9

gas emissions resulting from human activities. These human-generated gases derive from aspects such as building construction and operation, land-use planning, and transportation systems and infrastructure [2]. Climate events place a great burden on health systems such that responsibility that lies with health systems in the face of climate change is enormous. For this purpose, strengthening public health services must be a central component of all *nations'* climate change adaptation measures and policies. The intensification of these events has a direct or indirect effect on human health by disrupting ecosystems, agriculture, food, water quality and availability, air quality, and damaging infrastructure [3]. The direct effects of global warming and climatic change are heat-related illnesses, infectious diseases, cardio-vascular diseases, injuries, and respiratory diseases leading to premature deaths. Due to the fact that increasing GHGs threaten the environment and healthcare, authorities encounter considerable challenges to reduce carbon footprint in health systems. To help in decision-making for both cost and carbon in healthcare, Pollard et al. [4] proposed a bottom-up modeling framework. Research findings confirm that a bottom-up model is an efficient tool in the process of estimating and modeling the carbon footprint of healthcare. Emergency department (ED) is a primary healthcare department, the main entrance to the hospital, and a key component of the healthcare system. EDs are overcrowded and the length of stay (LOS) of patients has increased, whereas the quality of service has decreased. The overcrowding of EDs is a worldwide issue and a national crisis in the USA [5]. Many researches have been done on ED overcrowding since 20 years ago [6]. Despite that EDs are under this overcrowding phenomenon, they suffered from budget reductions. Therefore, new techniques should be found in order to deal with such an overcrowded condition. ED managers require different and fresh solutions because society demands not only care, quality, and service but also the best care, quality, and service. A direct solution to this issue is increasing the size of EDs. However, this straightforward solution is limited by facility, number of staff and services, and it is not the best approach. Also, healthcare managers have to maximize the use of healthcare resources, i.e. to optimize the performance of the ED in order to increase the satisfaction of the patients. Nevertheless, the increasing use of healthcare resources leads to an increasing carbon footprint in the patient flow. Cabrera et al. [7] presented an agent-based modeling and simulation to design a decision support system for the operations of EDs. The aim of this chapter is to find the optimal ED staff configuration, which consists of doctors, triage nurses, and admission personnel. For this, two different criteria, to minimize patient waiting time and to maximize patient throughput, were proposed and tested. Solutions obtained by applying an exhaustive search technique yield promising results and a better understanding of the problem. Results of this research enable health service providers to provide a more reliable reference for their decisions. Zhao and Lie [8] proposed different discrete event models of ED to enable managers to predict the future number of resident patient in each department/ward. Bhattacharjee et al. [9] classified the existing approaches related to modeling patient flows in hospitals focusing on the recent advancements to identify future research avenues. They proved a generic framework for patient flow modeling and performance analysis of hospital systems. Zhecheng [10] proposed and developed an online prediction procedure based on discrete-event simulation to predict the bed occupancy rate in a short term period. Simulation results showed that the predicted values were closer to the actual values with a narrower confidence interval compared to the offline approach. Zhu et al. [11] presented a discrete-event simulation (DES) model to help the healthcare service providers determine the proper ICU bed capacity which strikes the balance between service level and cost-effectiveness. The simulation results and the actual situation show that the DES model accurately captures the variations in the system, and the DES model is flexible to simulate various what-if scenarios. There are a wide variety of indicators to improve service delivery and patient satisfaction to evaluate the effectiveness of different parts of the hospital. Among these indicators, *patients'* length of stay, waiting time, and number of discharges are of particular importance. In addition, by changing the *community's* expectations of the hospital, patients are no longer willing to tolerate poor quality treatment and their look at the system is like that of a customer, so the concept of service has shifted from optimizing the use of resources to finding a balance between quality of service to patients and operational efficiency for health providers. Therefore, while reducing the number of patients lost, patient satisfaction should be increased by increasing the quality of their treatment and reducing waiting times, each of which is costly and should be treated with caution. The previous research focus has been addressed towards classical-based objectives while ignoring environmental-based objectives. In his research is considered an environmental-based objective besides classical objectives. The remainder of the chapter is organized as follows: in the next section green approaches in healthcare are reviewed, Sect. 9.3 introduces emergency department, and Sect. 9.4 proposed case study, Sect. 9.5 proposed the research methodology and simulation model. Section 9.6 presents the OptQuest optimization method of the simulated model and its experiments. The discussion and conclusions that can be drawn from our analysis do exist in Sect. 9.7.

9.2   Green Approaches in the Healthcare

The Green Guide for Healthcare [2] identifies opportunities to enhance environmental performance in the following domains: site selection, water conservation, energy efficiency, recycled and renewable materials, low-emitting materials, alternative transportation, daylighting (the use of natural light in a space to reduce electric lighting and energy costs), reduced waste generation, local and organic food use, and green cleaning materials. Some decisions, such as site selection, occur during the planning and construction phases; other decisions, such as food sourcing and cleaning practices, are primarily questions of operation after a building is completed. Commitments to energy conservation, renewable resource use, and similar principles must be made and reinforced throughout the life cycle of a facility, from building conception through operation and replacement. According to [12] use of green technologies to reduce CO_2 can be categorized as an environmentally oriented sustainability action. The consumption and generation of energy are associated with significant damages for the climate, the environment and, consequently, the economy. Greenhouse gas emissions have a decisive factor in climate change and global warming. The delivery of healthcare services produces remarkable greenhouse gas emissions. Some researchers have addressed the environmental effect of medical gases. To illustrate, Gilliam et al. [13] estimated direct CO_2 emissions from laparoscopic surgeries. In another research, Ryan and Nielsen [14] determined the 20-year global warming potentials of three common anesthetic gases including sevoflurane, isoflurane, and desflurane. As a consequence, they applied these gases to clinical scenarios in order to estimate the impacts of them on the environment. According to climatological reports of the National Oceanic and Atmospheric Administration (NOAA) in 2018, human activities have been known as the main causes of the increase in global temperature. Carbon dioxide and methane are among the GHGs that have the dominant impact on warming up and temperature rise-up on the earth. GHGs directly affect human health through both diffusions of diseases and pandemics and food shortages around the globe [15]. Based on the United Nation's Food and Agriculture Organization reports, climate change has decreased agricultural production. World Health Organization claims that pollution, diseases, such as malaria and cholera that spread in most areas of the planet, and severe heat causes cardiovascular disease are of GHGs effects. In 2013, the US healthcare sector was announced to be responsible for significant fractions of air pollution emissions and impacts, including acid rain (12%), greenhouse gas emissions (10%), smog formation (10%), criteria air pollutants (9%), stratospheric ozone depletion (1%), and carcinogenic and non-carcinogenic air toxics (1–2%). Due to the importance of complicated equipment and technologies at the process of treatment in health systems especially hospitals, CO_2 emission must be managed in direct and indirect manners [16, 17].

Healthcare system like other industries has been required to improve its environmental performance in past decades, and many new healthcare facilities have been built worldwide based on this view [18]. The amount of GHGs emitted by the use of electrical equipment in the *hospital's* intensive care unit is calculated by Pollard et al. [19] with a precise estimation of the power consumption. Given that climate change is one of the most important problems related to public health, so reducing greenhouse gas emissions from health care is one of the key responsibilities of the health system in preventing global warming. Considering the importance of *patient's* flow into emergency departments on occupying medical equipment that needs consuming electrical power for running, the patient flow has been studied in Bushehr Heart Hospital (BHH). Since this study attempts to estimate carbon dioxide produced by electric equipment in a health center in order to preserve the environment and reduce the trend of carbon increase, it could be a significant step towards managing produced CO_2 and small steps to prevent it.

9.3 Emergency Department

The ED is the first point of access for many hospital visits and a major source of unplanned episodes of patient care. EDs are open and staffed 24 h per day, 365 days per year, including holidays. The main functions of the ED include treatment of the acutely unstable and unwell and also those with nonlife-threatening conditions requiring immediate attention. EDs are large, complex, and dynamic units that have to increase their resources to attend all of those cases. The process begins by entering patients to ED by random or on-intermittent in two forms. Directly is the patient referring to ED by itself or using ambulance. Vital signals of the patient will be controlled at triage, and nurse decide whether patient needs hospitalization or not. If not, then process is over, and if yes the process will be continued. Based on the acuteness of the patient's condition and the nurse's opinion, the patient will be classified so that to determine the level of necessary treatment [20]. ED classifies patients into five levels. Level 1 that is in a less percent are patients who do not have often awareness and are at the highest emergency level for surviving their life. Level 2 are patients with dangerous signals and severe pain and peril of falling into coma, thus they cannot wait, level 3 are patients with two or more emergency treatment actions, level 4 needs one treatment action, and finally in level 5 are patients who do not have demand for instantaneous treatment and can wait. This class of referrals refers to a located clinic in the hospital by nurses.

9.4 Case Study

Our case study is the emergency department of Bushehr Heart Hospital, in Southern Iran. The data gathered from August 2016 to August 2017 covers all patients who have visited the hospital during this period. The total number of patients who referred to the hospital was almost 7807. About (5%) of them diagnosed to have ESI1, (5%) for ESI2, (30%) for ESI3, (30%) for ESI4, and (30%) of patients categorized to be ESI5. Using patient's arrival records time in triage database the distribution of interarrival times founded to be an exponential distribution with average 67 min (EXPO (67)). In doing this, a sample of the obtained data is tested by Chi-Square(χ^2) and Kolmogorov–Smirnov test in input analyzer arena and SPSS software. According to Table 9.1, the distribution of *patients'* arrival to triage is considered to follow an exponential distribution.

The other required data is the occupation of each equipment which has been used in the treatment process, to doing this some cited statistical procedure has

Table 9.1 The goodness of fit results of patients arrival pattern

Arrive to ED	Sample	Test type	Statistical test	p-value	Distribution
	150	χ^2	0.723	0.423	Exponential

Table 9.2 Information on medical equipment in ED

| Department | | Medical equipment | NEQ | DTEQ | | | CR |
				Min	Ave.	Max	
ED	Triage	Monitoring of vital signs	1	1	2	3	0.14
	CPR	CPR beds	1	45	82	120	2.64
		Monitoring of vital signs	1	45	60	90	0.14
		Syringe pump	3	30	50	75	0.025
		General motorized suction	2	20	30	40	0.15
		ElectroShock	2	15	20	30	0.22
	IED	Portable ventilator	1	10	50	80	0.529
		Blood gas analyzer	1	5	7	10	0.016
		IED beds	7	240	300	360	0.5
		Monitoring of vital signs	7	240	300	360	0.14
		Syringe pump	7	240	300	360	0.025
		Electrocardiograph	7	3	4	5	0.14
		Echocardiograph	2	15	17	20	1

Note. NEQ, DIEQ, and CR stand for number of electrical equipment, demanding time to use electrical equipment per patient (triangular distribution), and constant rates of electricity consumption (kilowatt-hours)

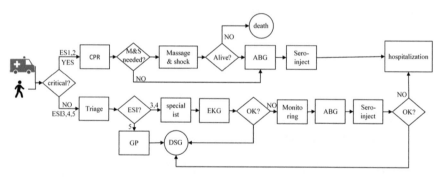

Fig. 9.1 Patient flow in the ED

been applied and the results show that equipment occupation time has triangular distribution with parameters and for each device. Table 9.2 presents the set of medical equipment and time distribution used for treatment processes. Simulation of CFP is performed for all units in 1 year, 2 days warm-up period with 50 runs and 10 replications. Warm-up period is set for simulation run to eliminate any bias at the early stages of the process.

Simulation is performed for all units of ED in one year, two days warm-up period with 10 runs. The warm-up period is set for simulation run to eliminate any bias at the early stages of the process. The described patient flow in ED and its flowchart is shown in Fig. 9.1. Following the process patients in levels 1 and 2 refer to the CPR department; 3, 4, and 5 refer to the medical visit room and can be discharged if they had stable and normal conditions, by prescribing medications, otherwise medical tests, hospitalization, or radiology, based on specialist doctors. There is a probability for patients to revisit doctors based on test results and demanded treatment process. Finally, doctors decide for prescribing, discharging, or hospitalizing patients in unit, or in CCU, ICU, PCCU, or surgery room; thus, they exit from ED. As illustrated, patients may come on foot or by ambulance and according to his ESI, he will be routed to either CPR or the triage. Following the care procedure, a patient may leave the ED, as he does discharge (DSG) or being hospitalized for further treatments. Also, in some cases a patient dies (the "death" symbol in the illustration).

9.5 Research Methodology and Simulation Model

Simulation can be used as an effective analysis technique to create, maintain, evaluate, or improve a system or process [21]. The ED of the hospital has three care units (workstations) including triage, cardiopulmonary resuscitation (CPR), and inpatient emergency department (IED). In this research, we take the concept of flexible job-shop scheduling (FJSP) in modeling the ED care processes. For this, we consider patients as jobs, treatment processes as operations, and medical equipment as machines. The main goal of this research is to calculate and optimize the CFP in the *patient's* flow, so medical equipment and the duration that is used of this equipment at each care unit are considered in the treatment process. To making the decision for patients flow in the studied hospital, discrete-event simulation (DES) has been applied for this purpose, the Arena14.0 package has been applied to simulate the system. The steps of the proposed simulation-optimization framework are inspired by Uriarte et al. [21]. It is based on five steps, understanding and data collection, simulation modeling, selection of performance criteria, multi-objective optimization, decision-making. In the first step studying the intended system needs sufficient comprehension of process and systems performance. Therefore, in this step, patients' behavior in the treatment process will be investigated in different units of the hospital, demanded data will be gathered, and then a design of a conceptual model of patient flow. In the second step, a model of the current situation has been simulated based on gathered data and conceptual model of patient flow in the studied system and finally at this step after confirmation and validation of the simulation model, a computerized model is presented. In the third step, the model needs criteria performance for evaluating and comparing the system performance. At this step key performance criteria of the hospital are collected. In the fourth step after creating multi-objective optimization model, simulated model by OptQuest is resolved using e-constraint technique. Finally, at the fifth step some recommendations are presented for improving the responsibility of healthcare system based on the obtained results.

Table 9.3 Model validation result. Total number throughput patients

Num.	Simulation	Real data	t-Statistics	Sig.
Patient output	7326	7328	0.561	0.605

Table 9.4 Model validation results. Length of stay (min.)

Unit care	Simulation (min.)	Real data (min.)	t-Statistics	Sig.
Triage	2.0512	2	−0.629	0.564
IED	389.24	390	−0.351	0.743
CPR	85	80	−4.767	0.09

9.5.1 Verification and Validation of Simulation Model

The simulation model of the current conditions approved by the executive manager and validation has been used to show that the simulation model is an exact representation of the real-world system. Results from validation are shown in Tables 9.3 and 9.4.

9.5.2 Carbon Footprint Calculation

The emission factor (EF) expressed in terms of the mass of carbon (or carbon dioxide) emitted for every unit of energy is delivered. The emission factor (EF) of the country obtained through Eq. (9.1) is as follows:

$$EF = kgCO2/kWh \tag{9.1}$$

where $kgCO_2$ and kWh indicate kilograms of CO_2 and kilowatt electricity used per hour, respectively. The total amount of produced $kgCO_2$ in the ED (T_{CO_2}) is calculated as follows:

$$T_{CO_2} = \sum_{i=1}^{I} \sum_{j=1}^{J} \sum_{k=1}^{K} (EF * CT_{ijk} * W_{jk} * Z_{ijk}) \tag{9.2}$$

where the related variables and parameters are

EF: Emission factor
CT_{ijk} : Usage time (hour) of equipment j in care unit k per patient i
W_{jk} : Power consumption rate of equipment j in care unit k (kW)
Z_{ijk} : If equipment j is used in care unit k for patient i 1; otherwise, it is zero.

Table 9.5 Average CFP produced in ED for one year	Care units		CFP (kg)
	Emergency department	Triage	7
		CPR	119
		Inpatient ED	17670
	Total CFP		17796

9.5.3   Simulation Outputs of Hospital Performance

Average results from simulating the status are summarized in Table 9.5. CFP column shows the average produced carbon dioxide for each hospitalized patient.

9.6   Simulation Optimization

Simulation optimization is a modeling method applied for problems that objective function or some of the constraints can only be evaluated by simulation, for example, optimization model includes functions that are not able to be evaluated analytically. Simulation optimization purpose is gaining the best framework of system to attain a specific goal with scarce resources on the one hand and on the other one, while conducting real-world systems, evaluation some of them is highly time taking. Considering simulation results for CFP in patient's flow which is not totally optimized, OptQuest has been used.

9.6.1   OptQuest

OptQuest is an optimization software embedded in simulation packages like Arena. It increases the analysis abilities by providing searching optimized solutions. In fact, OptQuest is a simulation-optimization engine, built on a unique set of powerful algorithms and sophisticated analysis techniques including Tabu-search, neural networks, and scatter search meta-heuristic algorithms. OptQuest searches, adjusts, and analyzes input values and identifies the best possible outcomes with unparalleled efficiency. This method has been applied to studying the treatment system [22–25]. In this chapter, OptQuest has been used to estimate the optimal control levels for optimizing ED simulation model.

9.6.2 Optimization Model

In this study, four objective functions are evaluated, that is, the minimum produced CO_2 (f_1), the minimum length of stay in ED (f_2), the minimum waiting time (f_3), and the maximum number of patient being fully treated (f_4). The mathematical model of the multi-objectives is proposed as follows. Then, OptQuest is used to obtain an optimized framework of ED resources, for more hospital responsibility (decrease in waiting time and LOS and increase in patient throughput), and reduces in produced CO_2 amount.

$$Min f_1(x_1, x_2, x_3, x_4, x_5, x_6, x_7, x_8, x_9, x_{10}, x_{11}, x_{12}) \tag{9.3}$$

$$f_2(x_1, x_2, x_3, x_4, x_5, x_6, x_7, x_8, x_9, x_{10}, x_{11}, x_{12}) \leq 310 \tag{9.4}$$

$$f_3(x_1, x_2, x_3, x_4, x_5, x_6, x_7, x_8, x_9, x_{10}, x_{11}, x_{12}) \leq 4.51 \tag{9.5}$$

$$f_4(x_1, x_2, x_3, x_4, x_5, x_6, x_7, x_8, x_9, x_{10}, x_{11}, x_{12}) \leq 7809 \tag{9.6}$$

$$x_{1L} \geq x_1 \leq x_{1U} \tag{9.7}$$

$$x_{2L} \geq x_2 \leq x_{2U} \tag{9.8}$$

$$x_{3L} \geq x_3 \leq x_{3U} \tag{9.9}$$

$$x_{4L} \geq x_4 \leq x_{4U} \tag{9.10}$$

$$x_{5L} \geq x_5 \leq x_{5U} \tag{9.11}$$

$$x_{6L} \geq x_6 \leq x_{6U} \tag{9.12}$$

$$x_{7L} \geq x_7 \leq x_{7U} \tag{9.13}$$

$$x_{8L} \geq x_8 \leq x_{8U} \tag{9.14}$$

$$x_{9L} \geq x_9 \leq x_{9U} \tag{9.15}$$

$$x_{10L} \geq x_{10} \leq x_{10U} \tag{9.16}$$

$$x_{11L} \geq x_{11} \leq x_{10U} \tag{9.17}$$

$$x_{12L} \geq x_{12} \leq x_{10U} \tag{9.18}$$

$$x_1, x_2, x_3, x_4, x_5, x_6, x_7, x_8, x_9, x_{10}, x_{11}, x_{12} \in Z^+ \tag{9.19}$$

In this optimization model, the amount of CO_2 produced in the patient flow (Eq. (9.3)) has been considered as an objective function. f_1, f_2, f_3, and f_4 are functions of the vector x_i representing decision variables including medical equipment in ED. Equations (9.3), (9.4), and (9.5) display constraints, where f_2, f_3, and f_4 are the minimum length of stay, the minimum waiting time, and increase patient throughput, respectively. The right-hand side of constraints (9.4)–(9.6) have been found using simulation. Equations (9.7)–(9.18) are the number of medical equipment in ED that is determined by hospital management and supervisors. In

Table 9.6 Bounds of variables

Variable	Variable name	Low	Up
x_1	ElectroShock in CPR	1	3
x_2	Hospital bed in CPR	1	2
x_3	Monitoring of vital signs in CPR	1	2
x_4	General motorized suction in CPR	1	3
x_5	Portable ventilator in CPR	1	2
x_6	Syringe pump in CPR	2	5
x_7	Electrocardiograph	1	3
x_8	Echocardiograph	1	2
x_9	Hospital bed in IED	6	9
x_{10}	Monitoring of vital signs in IED	6	9
x_{11}	Syringe Pump in IED	6	9
x_{12}	Monitoring of vital signs in triage	1	2

addition, f_2, f_3, and f_4 are randomized functions. That is, they do not have any analytical form and they can be only evaluated by simulation. Equation (9.19) indicates that the mathematical model is an integer problem. Table 9.6 presents the lower and upper bounds of variables. Table 9.6 presents the low and up bounds of variables $x_1, x_2, \ldots, x_6, x_7, x_8, x_9, x_{10}, x_{11}, x_{12}$.

To solve the model, while having multi-objective function, one of the objectives will be set as the objective function, while the others set to be constraints. The right-hand side of those constraints should be carefully set, to ensure that the optimal solution to the SOOP guarantees the MOOP solution. Since the OptQuest implemented for single-objective optimization problems, the Epsilon-Constraint (ϵ-*constraint*) technique [26] has been used to transform the multi-objective optimization problem (MOOP) into a single-objective optimization problem (SOOP). The OptQuest is implemented for all optimization problems with 10 runs and 10 repetitions. Four objective functions f_1, f_2, f_3, and f_4 cannot be met simultaneously (Table 9.7), because CO_2 minimization constraint in treatment process is in conflict with the increase in patients throughput.

Table 9.8 shows the OptQuest output for solving single- and multi-objective problems. According to Table 9.8, comparing results from both simulation and the OptQuest f_1, f_2, f_3, and f_4 have been improved by 0.89%, 0.97%, 24%, and 1.13%, respectively.

Table 9.7 Results of OptQuest for minimizing f_1 by considering f_2, f_3, and f_4 as constraints

Solution status	Solution	Configuration (number of medical equipment)											
		x_1	x_2	x_3	x_4	x_5	x_6	x_7	x_8	x_9	x_{10}	x_{11}	x_{12}
Infeasible	17793.78	2	1	1	2	1	3	2	1	7	7	7	1
Infeasible	17653.52	2	2	2	2	2	4	2	2	8	8	8	2
Infeasible	17616.66	2	2	2	2	2	5	2	2	9	9	9	2
Infeasible	17729.16	1	1	1	1	1	2	1	1	6	6	6	1
Infeasible	17616.66	3	2	2	3	2	5	3	2	9	9	9	2
Infeasible	17653.52	3	2	2	3	2	4	3	2	8	8	8	2
Infeasible	17616.66	3	1	1	2	2	2	3	1	9	9	8	2
Infeasible	17729.16	2	2	2	3	1	5	1	2	6	7	6	1
Infeasible	17729.16	3	2	1	3	2	5	1	1	6	8	8	1
Infeasible	17729.16	1	2	2	1	2	4	3	1	6	8	9	1

9.7 Conclusions and Future Insights

The CO_2 accounting stems from the equipment in care units is a basic step for greenness and lowering climate change rates. Therefore, simulating the produced CO_2 proportionally to the consumption amount of electricity by using a discrete-event simulation approach aims to control and optimize the efficiency of equipment. By the knowledge that the simulation is not an optimization method in itself, the OptQuest optimization method has been used for optimizing cardiovascular patients flow, based on multi-objectives such as minimizing the produced amount of CO_2, waiting time, length of stay, and increasing patient throughput. A comparison of the outputs obtained from the current position and OptQuest optimization confirms that this technique is efficient for optimizing the *patient's* flow and decreases produced CO_2.

The production of carbon footprint in hospitals is related to energy consumption and for the reason to make it more realistic, it can be a good idea if it is considered in material based. Discrete-event methodology in this chapter has been used and based on its performance for solving this kind of problems, to enhance the performance of the simulation; agent-based methodology can be applied. Also by the knowledge of restrictions of OptQuest for multi-objective optimization, it would be better to apply a meta-heuristic algorithm to reduce CO_2 and waiting time.

Table 9.8 Results of OptQuest for single/multi-objective optimization problems

Objective function		Constraint (s)	SS	Answer	Configuration (number of medical equipment)											
					x_1	x_2	x_3	x_4	x_5	x_6	x_7	x_8	x_9	x_{10}	x_{11}	x_{12}
One OBJ.	f_1	–	F	17636.86	2	2	2	2	2	5	2	2	9	9	9	2
	f_2	–	F	307	1	1	1	2	1	3	1	2	7	7	7	1
	f_3	–	F	3.41	2	2	2	3	2	5	2	3	9	9	9	2
	f_4	–	F	7897	1	2	1	3	1	5	2	2	6	8	6	2
Two OBJs.	f_1	f_2	F	17616.66	2	2	2	2	2	5	2	2	9	9	9	2
	f_1	f_3	F	17616.66	2	2	2	2	2	5	2	2	9	9	9	2
	f_1	f_4	NF	17729.16	1	1	1	1	1	2	1	1	6	6	6	1
	f_2	f_1	F	307.127	2	2	2	2	2	5	2	2	9	9	9	2
	f_2	f_3	F	306.97	2	2	2	3	2	5	2	3	9	9	9	2
	f_2	f_4	NF	307.27	2	1	2	2	1	5	2	1	9	7	6	1
	f_3	f_1	NF	3.41	2	2	2	2	2	5	2	2	9	9	9	2
	f_3	f_2	F	3.44	2	2	2	3	2	5	2	3	9	9	9	2
	f_3	f_4	NF	3.63	2	1	2	2	1	5	2	1	9	7	6	1
	f_4	f_1	NF	7836	2	2	2	2	2	5	2	2	9	9	9	2
	f_4	f_2	F	7814	2	2	2	3	2	5	2	3	9	9	9	2
	f_4	f_3	NF	7836	2	2	2	2	2	5	2	2	9	9	9	2
Three OBJs.	f_1	f_2, f_3	F	17616.66	2	2	2	2	2	5	2	2	9	9	9	2
	f_1	f_2, f_4	NF	17653.52	2	2	2	2	2	4	2	2	8	8	8	2
	f_1	f_3, f_4	NF	17729.16	1	1	1	1	1	2	1	1	6	6	6	1
	f_2	f_1, f_3	NF	307.127	2	2	2	3	2	5	2	3	9	9	9	2
	f_2	f_1, f_4	NF	307.127	2	2	2	3	2	5	2	3	9	9	9	2
	f_2	f_3, f_4	NF	307.127	2	2	2	3	2	5	2	3	9	9	9	2
	f_3	f_1, f_2	NF	3.44	2	2	2	3	2	5	2	3	9	9	9	2
	f_3	f_1, f_4	NF	3.411	2	2	2	3	2	5	2	2	9	9	9	2
	f_3	f_2, f_4	NF	3.63	2	1	2	2	1	5	2	1	9	7	6	1
	f_4	f_1, f_2	NF	7814	2	2	2	3	2	5	2	3	9	9	9	2
	f_4	f_1, f_3	F	7836.4	2	2	2	2	2	5	2	2	9	9	9	2
	f_4	f_2, f_3	F	7836.4	2	2	2	2	2	5	2	2	9	9	9	2
Four OBJs.	f_1	f_2, f_3, f_4	NF	17729.16	1	2	2	1	2	4	3	1	6	8	9	1
	f_2	f_1, f_3, f_4	NF	307.127	2	2	2	3	2	5	2	3	9	9	9	2
	f_3	f_1, f_2, f_4	NF	3.44	2	2	2	3	2	5	2	3	9	9	9	2
	f_4	f_1, f_2, f_3	NF	7814	2	2	2	3	3	5	2	2	9	9	9	2

Note. SS, F, NF, OBJ stand for solution status, feasible, infeasible, and objective, respectively

References

1. K. Fang, R. Heijungs, The role of impact characterization in carbon footprinting, Frontiers in Ecology and the Environment **13**(3), 130 (2015)
2. M. Younger, H.R. Morrow-Almeida, S.M. Vindigni, A.L. Dannenberg, The built environment, climate change, and health: opportunities for co-benefits, American journal of preventive medicine **35**(5), 517 (2008)
3. W.H. Organization, et al. Quantitative risk assessment of the effects of climate change on selected causes of death, 2030s and 2050s, (2014)
4. A.S. Pollard, T.J. Taylor, L.E. Fleming, W. Stahl-Timmins, M.H. Depledge, N.J. Osborne, Mainstreaming carbon management in healthcare systems: A bottom-up modeling approach, Environmental science & technology **47**(2), 678 (2013)
5. I. of Medicine Committee on the Future of Emergency Care in the US Health System, et al. Hospital-based emergency care: at the breaking point, Hospital-based emergency care: at the breaking point (2006)
6. S.G. Lynn, A.L. Kellermann, Critical decision making: managing the emergency department in an overcrowded hospital, Annals of emergency medicine **20**(3), 287 (1991)
7. E. Cabrera, M. Taboada, M.L. Iglesias, F. Epelde, E. Luque, Optimization of healthcare emergency departments by agent-based simulation, Procedia computer science **4**, 1880 (2011)
8. L. Zhao, B. Lie, Modeling and simulation of patient flow in hospitals for resource utilization, in *Proceedings of the 49th Scandinavian Conference on Simulation and Modeling* (2008)
9. P. Bhattacharjee, P.K. Ray, Patient flow modelling and performance analysis of healthcare delivery processes in hospitals: A review and reflections, Computers & Industrial Engineering **78**, 299 (2014)
10. Z. Zhecheng, An online short-term bed occupancy rate prediction procedure based on discrete event simulation, Journal of Hospital Administration **3**(4), 37 (2014)
11. Z. Zhu, B. Hoon Hen, K. Liang Teow, Estimating ICU bed capacity using discrete event simulation, International journal of health care quality assurance **25**(2), 134 (2012)
12. M. Marimuthu, H. Paulose, Emergence of sustainability based approaches in healthcare: Expanding research and practice, Procedia-Social and Behavioral Sciences **224**, 554 (2016)
13. A. Gilliam, B. Davidson, J. Guest, The carbon footprint of laparoscopic surgery: should we offset? Surgical endoscopy **22**(2), 573 (2008)
14. S.M. Ryan, C.J. Nielsen, Global warming potential of inhaled anesthetics: application to clinical use, Anesthesia & Analgesia **111**(1), 92 (2010)
15. J. Hartter, L.C. Hamilton, A.E. Boag, F.R. Stevens, M.J. Ducey, N.D. Christoffersen, P.T. Oester, M.W. Palace, Does it matter if people think climate change is human caused? Climate Services **10**, 53 (2018)
16. A.J. MacNeill, R. Lillywhite, C.J. Brown, The impact of surgery on global climate: a carbon footprinting study of operating theatres in three health systems, The Lancet Planetary Health **1**(9), e381 (2017)
17. L.H. Brown, P.G. Buettner, D.V. Canyon, The energy burden and environmental impact of health services, American journal of public health **102**(12), e76 (2012)
18. M. Pinzone, E. Lettieri, C. Masella, Sustainability in healthcare: Combining organizational and architectural levers, International Journal of Engineering Business Management **4**, 38 (2012)
19. A. Pollard, J. Paddle, T. Taylor, A. Tillyard, The carbon footprint of acute care: how energy intensive is critical care? Public health **128**(9), 771 (2014)
20. S. Saghafian, W.J. Hopp, M.P. Van Oyen, J.S. Desmond, S.L. Kronick, Patient streaming as a mechanism for improving responsiveness in emergency departments, Operations Research **60**(5), 1080 (2012)
21. A.G. Uriarte, E.R. Zúñiga, M.U. Moris, A.H. Ng, How can decision makers be supported in the improvement of an emergency department? A simulation, optimization and data mining approach, Operations Research for Health Care **15**, 102 (2017)

22. J.P. Kleijnen, J. Wan, Optimization of simulated systems: OptQuest and alternatives, Simulation Modelling Practice and Theory **15**(3), 354 (2007)
23. J. April, F. Glover, J. Kelly, M. Laguna, Simulation/optimization using "real-world" applications, in *Proceeding of the 2001 Winter Simulation Conference (Cat. No. 01CH37304)*, vol. 1 (IEEE, 2001), vol. 1, pp. 134–138
24. M.C. Fu, F.W. Glover, J. April, Simulation optimization: a review, new developments, and applications, in *Proceedings of the Winter Simulation Conference, 2005.* (IEEE, 2005), pp. 13–pp
25. F. Glover, J.P. Kelly, M. Laguna, The OptQuest approach to Crystal Ball simulation optimization, Graduate School of Business. University of Colorado (1998)
26. Y.H. YV, L.S. Lasdon, D. DA WISMER, On a bicriterion formation of the problems of integrated system identification and system optimization, IEEE Transactions on Systems, Man and Cybernetics (3), 296 (1971)

Chapter 10
The Effect of Risks on Discrete Event Simulation in Healthcare Systems

Milad Poursoltan, Malek Masmoudi, and Mamadou Kaba Traore

Abstract A wide variety of projects are performed in healthcare in order to improve managerial decisions for efficient allocation of material, staff, and financial resources. However, traditional approaches do not take into account the inevitable process variability, uncertainty, scale, and interconnections that are critical for making efficient managerial decisions in real healthcare systems. Thus, engineers are increasingly involved in using computer simulation which is able to capture all these factors into efficient managerial decision-making. But, when compared to other applications, the simulation project in healthcare is more likely to fail and has a poor success rate. This chapter draws on the discrete event simulation (DES) project in healthcare and discusses the agenda of DES projects in healthcare focusing on (1) a characterization of DES project and simulation process, (2) risks on DES project in healthcare, and (3) applicable risk analysis methods. This chapter contains a real case study—an emergency department—which seems to be the most popular area of DES projects in healthcare.

10.1 Motivation

DES is known as a decision-making tool which allows administrators, engineers, and managers to analyze the impact of operational and managerial decisions [1]. Due to the different functionalities of DES such as handling variability, complexity, and what-if analysis, its application in healthcare systems has gained more and more attention of researchers [2, 3].

M. Poursoltan (✉) · M. K. Traore
IMS UMR 5218, University of Bordeaux, Talence, France
e-mail: milad.poursoltan@u-bordeaux.fr; mamadou-kaba.traore@u-bordeaux.fr

M. Masmoudi
University of Jean Monnet Saint-Etienne, Sciences and Technologies Faculty, University of Lyon, Saint Étienne, France
e-mail: malek.masmoudi@univ-st-etienne.fr

© Springer Nature Switzerland AG 2021
M. Masmoudi et al. (eds.), *Operations Research and Simulation in Healthcare*, https://doi.org/10.1007/978-3-030-45223-0_10

According to [4], DES is the most powerful and versatile methodology for analyzing several problems related to healthcare systems such as capacity dimensioning, patient flow management, staffing assignment, and activities scheduling. However, at the project level, due to the features and environment of implementation, DES projects in healthcare are facing high rates of failure in different ways [5–8].

In spite of the fact that healthcare systems are evolved and changed over time, DES must be capable of making accurate predictions of the system under study in the expected time interval. However, when compared to other applications, the simulation project in healthcare is more likely to fail [7, 8]. Previous studies in the successfulness of healthcare simulation have pointed that the adaptation of standard approaches for developing and also using project management techniques are required to implement simulation projects successfully (e.g. [7, 9, 10]). However, despite the growing use of DES projects, very few researches have been performed to study the potential causes of failures. Emergency departments are perhaps the most-used components of healthcare system for DES project [6]. This chapter deals with DES project for emergency department receiving more than 23,000 patients per year. This case study illustrates the risks and challenges emerging in DES projects in practice.

10.2   Background

Computer simulation is an active domain of research which has been used to study the effect of a wide range of subjects in different systems, particularly in healthcare. Reference [6] classified papers according to the areas of application in healthcare. Reference [11] used a simulation model for an emergency department (ED) to provide a better understanding of the system behavior and improve the performance of ED operations. Reference [12] studied four different healthcare clinic layout scenarios by DES. Reference [13] designed a simulation model for a clinical environment and detailed the whole process, including the build, verification, validation, and test of procedures. A state of the art in healthcare systems modeling and simulation can be found in [14].

A number of authors have studied the successfulness of DES projects in healthcare. Reference [5] reviewed literature in healthcare simulation in the 1970s and found that less than 8% of them report the outcomes of successful implementation. According to [6], the obstacles to the successful implementation of healthcare simulation projects still exist to some degree. Reference [7] found that healthcare simulation has been less successful because there are significant differences across a range of factors in healthcare.

Several studies have been conducted based on the factors of success and failures in DES projects. Reference [15] points out that the user acceptance of simulation process is the key of project's success. Reference [16] argues that the data, the lack of understanding decision-makers, and the conflicting objectives of stakeholders are three main challenges for healthcare simulation success. Reference [17] proposes a top-down framework with 15 key performance indicators which are linked to a set of critical success factors that represent the level of success of the simulation projects. Reference [10] discusses ten key steps for successful simulation studies and believe that the technical knowledge and project management skills are necessary to develop a model that is valid for decision-making processes. Reference [9] provides a guideline including managerial skills that are needed in each step of the simulation process for the project success. Reference [18] investigates the characteristics of DES projects failures and success. Reference [19] describes success based on a four-stage model of changing perceptions and shows that success changes continuously throughout the life cycle of a DES project. Reference [20] discusses the key issues and challenges in healthcare modeling projects and proposes a framework towards successful implementation that divides a healthcare modeling project into three high-level components (pre-model, model, and post-model). Reference [21] links simulation success to implementation characteristics such as simulation software, operational cost, software environment, simulation output, organizational support, initial investment, and simulation task. Reference [22] characterizes successful DES projects by teamwork, cooperation, mentoring, effective communication of outputs, high quality vendor documentation, easily understood software syntax, higher levels of analyst experience, and structured approaches to model development. Reference [17] proposes five critical success factors: communication and inter-action, competence of the provider, responsiveness, involvement, and customer's organization.

The role of stakeholders in the successful implementation of healthcare simulation projects is highlighted in many studies (e.g. [2, 23–27]).

Some papers have discussed issues on specific subjects of success for DES projects. Reference [28] discusses statistical and probabilistic issues in simulation experiments. Reference [29] describes several forms of verification and validation (V&V) methods that should be employed throughout the life cycle of simulation study to increase accuracy and confidence in the model. From the perspective of project management, [30] argues patient flow improvement in healthcare and tries to describe management methods and techniques in this field.

Reference [31] points out that computer simulation studies should be seen as examples of information system projects. Risk management in software projects that present some similarities to risk management in DES project has been conducted in several studies (e.g. [32–38]).

10.3   Characterization, Uncertainty, and Simulation Process of DES Project in Healthcare

As a worldwide accepted reference for providing definitions, [39] describes uncertainty as the state, even partial, of deficiency of information about the understanding or knowledge of an event and its likelihood or consequence. In fact, not all uncertainties are risk but all risks are uncertain [40]. Accordingly, the risk for DES project is concerned with unpredictable events that might happen in the future, whereas their likelihoods and outcomes are uncertain and could potentially affect their objectives in some way. DES in healthcare lays emphasis on project, simulation, and healthcare characteristics that are the source of uncertainty. These characteristics are explained as follows:

1. Project characteristics: A range of features related to a project, such as deliverables provided, stakeholder influence, resources, constraints, and project process design, which inevitably generate uncertainties.
2. Simulation characteristics: Inherent characteristics of the simulation method such as prediction, statistical data, and estimation.
3. Healthcare characteristics: Common characteristics related to healthcare systems, such as central role of people, communication, challenges, and complexity.

The relations between the characteristics of DES project in healthcare are explained in Table 10.1.

To master uncertainties involved in DES, the simulation process needs to be understood. Different simulation life cycles and simulation process are used in literature, e.g. [41–43]. Despite emphasizing different aspects, Fig. 10.1 shows the simulation process that is considered in most of DES studies. The process initiates with the problem formulation. Reference [44] believes that this is the most important step in the life cycle of DES project. Once the decision for simulation is taken, an overall project plan should be prepared [45]. Objectives specify the questions that must be answered by the simulation. Then, the conceptual model should be provided. This model refers to the abstraction of the real-world system under study [46]. Abstraction emphasizes the need for simplification and assumptions of the real system [45]. Having good data to drive a model is just as important as having model logic and structure. Necessary and available data and how to gather them should be determined in the next step. Any inappropriate function in data collection affects the remaining project processes. After data collection, a computer simulation model is developed to convert the conceptual model into a digital format. Later, the verification ensures that the model does what it is intended to do and the validation performs the comparison of the model against real system to confirm a good real word representation. Once the model has been validated and verified, the scenarios and experiments should be designed. A simulation experiment refers to a test or series of tests in which modifications are made to the variables of simulation model [49]. Then, it is necessary to run the simulation for each scenario to have confidence in the results. Finally, the results should be documented. There are two types of documents in the model, one is related to the final report deliver to customer and the other is a complete set of documentation on the project itself [46].

10.4   Risks on DES Project in Healthcare

Risks are uncertainties that can affect project success. Over the last 50 years, the iron triangle of cost, time, and quality has been considered as the main success

Table 10.1 Description of characteristics that make a DES project in healthcare uncertain

Characteristics		Description
Project	Deliverables provided	Deliverables are tangible or intangible result of the project that is intended to be delivered to the client. DES in healthcare can provide an evidence-based tool to develop objectively operational solutions prior to implementation for decision-makers. The information is considered as a major deliverable in this kind of project
	Stakeholder influence	Stakeholder requirements can be varying, overlapping, and sometimes conflicting, leading to risks in project execution and acceptance. The variety of stakeholders in DES projects in healthcare is remarkable. Stakeholders can vary from strategic health authority managers to ambulance service clinicians with different expectations
	Resources	All projects are realized by resources, including people, material, tools, and money. All of these resources are unpredictable and add uncertainty to the projects in which they are involved
	Constraints	Assumptions and constraints may turn out to be wrong, and it is also likely that some will remain hidden or undisclosed. They present a source of uncertainty in the projects On DES projects in healthcare, the constraints and restrictions include project time, cost, accuracy level of outcome, level of information access, etc.
	Project process design	For each project, features and sequence of processes are designed according to project deliverables. Varieties of methods are applied for collecting data, validating, animating, and other project processes
Simulation in healthcare	Central role of people	A remarkable feature in health services is about people (managers, doctors, nurses, patients, etc.) and their behavior that affects healthcare systems. This characteristic will thread through every DES project in healthcare [47]
	Communication	There are cultural differences between healthcare staffs and simulation analysts and involve different ways of viewing problems and solving solutions. Reference [9] believes that poor communication is the single biggest reason for DES project failures
	Modeling challenges	A large number of challenges to DES projects in healthcare involve straight modeling. Healthcare systems are strongly dynamic. Models must be able to cover accurately healthcare characteristics
	Complexity	In healthcare systems, many components and interactions are not immediately visible to the observer. A large amount of time and effort may be required to describe the model with an appropriate level of detail.
Simulation	Statistical data	There is always a gap between statistical and real data. Conversely, random numbers are used in DES, and the use of a stochastic approach can be a subject of disconcertion as the results change slightly when running the model [48]
	Estimation	DES estimates the effects on the population, not on a particular person or event
	Future prediction	The prediction of the future is not possible. Therefore, there is an error level to predict the results of a system in the future

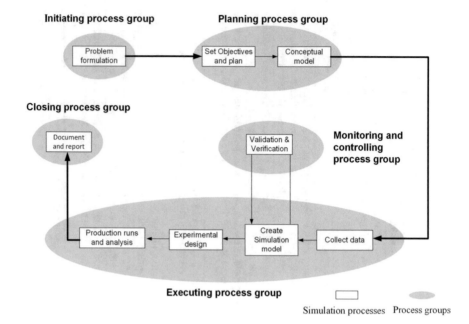

Fig. 10.1 Life cycle of DES project

factors of a project. The Project Management Institute (PMI) which is the largest professional organization dedicated to the project management field has published a guideline called PMBOK that is widely accepted. Reference [50] defines four factors of project success: quality, time, budget, and customer satisfaction. Quality factors in DES project have been studied by [19]. Reference [51] detailed the possible customer satisfaction factors in DES project. According to [52], the cost of DES project is considered by the costs of simulation software purchase and maintenance, hardware purchase and maintenance, simulation software training, training in simulation modeling, and human resources during the time of DES project required by customer.

10.4.1 Risk Identification

For DES projects, some risks have been mentioned in literature and named as challenges, errors, or obstacles (see Table. 10.2).

By studying the hundreds of common project risks that have been identified through various sources, we found that all DES project risks can be classified into six risk groups as follows: resources, planning, client's organizational environment, stakeholders, technical tasks, and system behavior (see Table 10.3). Our case study acknowledges the capability of this classification to cover all risks of DES project.

Table 10.2 Highlighted challenges for DES projects in literature

Challenges highlighted for DES projects in literature	Authors
The purpose of use, problem formation, and objectives	[5, 20, 44, 53]
Project management issues in DES project	[20–22, 30, 47, 54]
Complexity in healthcare systems	[47, 55]
Data issues	[20, 56]
Issues in simulation experiments	[28, 54, 56, 57]
Customers and stakeholder issues in simulation	[19, 20, 23, 24, 26, 27, 46, 58, 59]
Verification and validation of simulation models	[29, 41, 60, 61]
Modeling issues	[20, 44, 45, 47, 49, 56]

Except "system behavior" risk group which is defined especially for DES project in healthcare, the other risk groups exist in the literature with different forms. Since the healthcare system is always subject to change due to a large amount of variability related to people (e.g., doctors, nurses, patients, etc.), ignoring "system behavior" can lead to rework in projects or reduce the accuracy of the simulation results. We consider the risk in system behavior as those are related to a change in controllable parameters (CP), uncontrollable parameters (UP), or model layout, which leads to change in flow or system process. Any significant changes in the model layout or CPs can easily be detected, but most of the time UPs are discovered after comparing the expected results to the real results. In healthcare system, CPs are parameters such as number of beds, nurses, and equipment, working hours, etc. These parameters are considered to be controllable by healthcare managers, while UPs are parameters such as patient arrival times, treatment choice, patient characteristic, treatment duration, the rate of efficacy, cost, etc. These parameters in the model are believed to be fixed and uncontrollable in the current state. In most cases, UPs depend on a number of circumstances such as the location of hospital, types of served patients, types of available facilities, applied rules and regulations, etc. Any meaningful change in these parameters can change the behavior of healthcare system.

By the use of expert opinions as well as in literature, we found that four main factors may change the behavior of a healthcare system:

1. Trends: They have impacts on UPs. Reference [62] defines three trends for flow of patients: seasonal illness or incident, days of week, and patient arrival hours. In addition, processes that are newly established typically have trends in speed and quality. Therefore, they cannot be modeled correctly before stability.
2. Internal organization decisions: They are related to resources, like incentive programs, and related to processes, like the increase of service capacity or the achievement of new standards, i.e. a decision to open an air medical service would increase the rate of critical patient arrivals in an emergency department system. We noticed that internal organization decisions are not usually communicated to simulation specialists because they are unaware of the impact of their decisions on the simulation results.

Table 10.3 Risk classification for DES project in literature

Risk sources	Description
Resources	Human resources: Team composition of a DES project varies from one person playing many roles to a larger group of up to 10 or more playing single roles [52]. Risks related to human resources in a project cover team member turnover, staffing buildup, insufficient knowledge among team members, cooperation, motivation, and team communication issues [34]
	Material resources: DES projects need some kind of tangible work material such as hardware, software, and appropriate work environment. Uncertainty in material availability can bring risks into the project
Planning	Project planning is facing risks when project tasks and scheduling are not addressed properly. Usually excessive pressure or unrealistic scheduling increases project risk [34]
Client's organizational environment	The risks surrounding the organization under which the simulation is performed play a significant role in project success. It consists of several characteristics such as organizational maturity, organizational stability, organizational culture, decision hierarchy, organizational information systems, etc.
Stakeholders	Involvement of stakeholders in the modeling process is a key factor in successful model implementation [19, 20]. Reference [24] determined 19 organizational stakeholder groups to facilitate modeling in healthcare simulation
Technical tasks	The risks surrounding the organization under which the simulation is performed play a significant role in project success. It consists of several characteristics such as organizational maturity, organizational stability, organizational culture, decision hierarchy, organizational information systems, etc.
System behavior	The instability of the system, which is modeled in simulation projects, increases the risks and makes impacts on various input, process, and output parameters in a model. The four main factors behind the changes in system behavior have been particularly presented in this paper

3. External organization issues: These include credit availability, insurance services, social factors, political factors, technological factors, rules and regulations, etc. For example, a new vaccination law or a new government screening program can change medical process and patient flow in a hospital.
4. Simulations scenarios: Impact of simulation scenarios on system behavior is usually underestimated, while they can make hidden deviations in project results. For example, as nurses and physicians work in conjunction, by adding a new physician, maybe a new nurse should be added [62], or by adding doctors with

objective to reduce waiting time of patients in an emergency department system, we attract more patients and consequently impact waiting time.

The importance of each of these factors can vary from healthcare system to another. Thus, finding the importance of these factors can help the risk identification process to pay more attention to the factors that are more likely to lead to changes in the system behavior.

10.4.2   Risk Assessment

The risk assessment aims to find the magnitude of identified risks and then rank them. The selection of an approach to risk assessment depends on various factors such as availability of resources, nature and degree of uncertainty, available information, and complexity of application. To avoid wasting time and money in DES project in healthcare, we propose a two-stage risk analysis. Initial risk analysis should be performed by cost-effective tool such as failure mode and effect analysis (FMEA) for establishing risk priorities.

An additional risk analysis is carried out if identified risk is classified as a high risk by the initial risk analysis. In some cases, it may not have been possible to execute an additional risk analysis due to the lack of sufficient data to develop appropriate analysis. Reference [63] believes that a comparative semi-quantitative or qualitative ranking of risk by expert may still be effective in these circumstances. In order to fulfill the additional risk analysis with the appropriate degree of accuracy, two types of method are required to analyze the risks on model sensitivity and project execution in more detail. The VIKOR method and the event tree analysis (ETA) methods are able to assess the sensitivity of model and project execution to risks, respectively.

The ETA is a powerful tool for the evaluation and the quantitative analysis to determine potential consequences arising from a probable risk called initial event [64–66]. The magnitude of the initial event is the cumulative sum of the magnitude of all branches and the probability of occurrence of an event equals all probabilities of occurrence in the associated branch.

The sensitivity analysis is needed to determine how system target will be influenced by an event [67]. DES enables us to carry out the sensitivity analysis by testing different probable value of CPs and Ups (risks) for each scenario. One of the main goals of sensitivity analysis for risk assessment is to show the impacts of risks on the output results and scenarios priorities.

The VIKOR is a method of multi-criteria decision-making which has recently been used in simulation studies. Reference [68] considered this method for multi-response processes optimization, and [69] applied it to rank scenarios in simulation. We propose VIKOR to rank the scenarios based on the results of the sensitivity analysis. The scenarios are considered as feasible alternatives, key performance measures as criteria, and the probability of risk modes as weights of criteria.

10.5   Case Study

We studied the risks of DES project in an emergency department located in Tehran (Iran) that serves more than 23,000 patients per year. The objective of this DES project is to analyze four scenarios (defined by hospital managers) and determine the best one to improve the system.

The main output results of simulation projects in emergency departments are generally related to key performance measures (KPMs) that focus on lengths of stay, resource utilization, waiting time, left without being seen rate, cost services, or a set of these KPMs. In our project, waiting time and resource utilization are considered as two key performance factors with the same important coefficient according to managers' requirements. The risks are studied on cost, time, client satisfaction (understanding of the process, increasing decision speed), and quality (model, confidence in the model, data, competence of the provider, communication and interaction, involvement quality, responsiveness, customer's organization). Our emergency department model contains a number of UPs including the patient arrival time that allows generating entities and the characteristic of patient arrival that specifies the severity of patient condition and path of treatment.

To demonstrate the impact of each factor on the system behavior and indicate the more important factors, we prepared a questionnaire using the Likert approach to collect data from the emergency department staff. The reliability of the questionnaire was validated by Cronbach's alpha equal to 0.93. The results showed that 80% of the staff agreed that internal organization decisions can change the behavior of their emergency department system. For other factors such as trends, external organization issues, and simulations scenarios, the corresponding rates were 20%, 63%, and 33%, respectively.

To fulfill the initial risk analysis, we used FMEA with severity and likelihood 1–20 scales to allocate the risks into an appropriate class and a risk magnitude matrix was utilized for the purpose of finding the risk magnitude.

From initial risk analysis, we found some high risks in the project including three high risks on the system behavior as follows: change in number of patients (Risk 1), implementation of a new automation system (Risk 2), and change in insurance rules and checking out activities (Risk 3). Each risk could lead to change in several UPs in the model. The variation range on UPs for each risk (mode) was estimated by healthcare professionals in the emergency department. The conceptual model of our emergency department with high risks occurrence area (gray rectangles) and improvement scenarios is shown in Fig. 10.2.

Our simulation model is developed on Arena 15.0 which is one of the most used DES software. Figure 10.3 depicts our ED simulation model. This model contains four sections, namely reception, treatment, additional diagnosis, and discharge. Each section contains modules and each module has some UPs and CPs. This model calculates the average waiting time and the non-utilization rate of ED bed. In this model, the reception not only considers the number of patients, but also of their characteristics. After reception, there are three outputs for patient flow in the

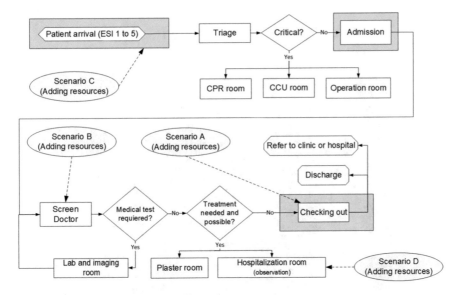

Fig. 10.2 Conceptual model, scenarios, and high risks occurrence areas

given department: discharge, death, or transfer to hospital. During the treatments, additional laboratory tests, radiology, and imaging may be needed. During the treatments, additional laboratory tests, radiology, and imaging may be needed.

The output results of the simulation model regarding 4 modes, current mode and three risk modes, are presented in Fig. 10.4 while considering four scenarios A, B, C, and D. The probabilities were assigned due to the experts.

We applied the VIKOR to rank the scenarios while taking the risks. Scenarios' results in Fig. 10.4 are considered as decision matrix then normalized in order to remove the units of decision matrix. As a minimization objective in this subject, the ideal and the worst solution are calculated. In the next step, the utility and the regret measure for each nondominated solution are obtained. finally, the VIKOR indexes are calculated.

The scenarios are ranked with regard to the best solution with the consideration of the risk modes. By comparing the results of prioritizing the scenarios in the presence and absence of risks, we found that scenario selection priorities are not affected and scenario A is the best choice. However, the expected results of key performance measures in this scenario can vary from 0.33 to 0.39 for non-utilization rate (ED bed) and 18.92 to 30.23 for average waiting time.

Even though the project team becomes sure that a model of simulation is valid, but this is no guarantee that stakeholders use it without being convinced of its benefits and its limitations [6, 15]. For high risks analysis in project execution, we use ETA. Figure 10.5 illustrates an ETA application for a high initial risk in our project. The probabilities were estimated by experts. We used a scale of 1–20 consequences

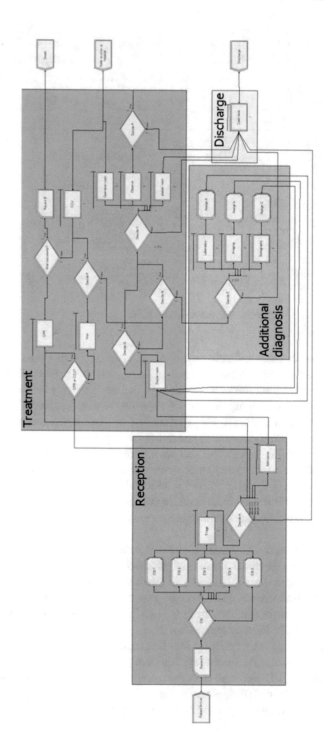

Fig. 10.3 Arena simulation model

Risk probability	0.45		0.35		0.15		0.05	
Mode	Current mode		Risk mode 1		Risk mode 2		Risk mode 3	
Weight of criterion	0.225	0.225	0.175	0.175	0.075	0.075	0.025	0.025
Scenario \ KPMs	Avarage waiting time (Min)	Nonutilization rate (ED Bed)	Avarage waiting time (Min)	Nonutilization rate (ED Bed)	Avarage waiting time (Min)	Nonutilization rate (ED Bed)	Avarage waiting time (Min)	Nonutilization rate (ED Bed)
Scenario A	28.29	0.33	30.23	0.33	20.92	0.39	18.92	0.39
Scenario B	34.56	0.38	22.64	0.36	31.19	0.39	11.46	0.41
Scenario C	30.24	0.34	33.18	0.36	20.57	0.33	34.44	0.35
Scenario D	41.73	0.29	32.57	0.36	22.70	0.37	20.84	0.38

Fig. 10.4 The results of scenarios in different modes

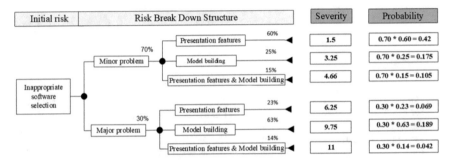

Fig. 10.5 An illustration for ETA application

scoring schema that represents the severity of risk. A combination of the severity of consequences and likelihood of occurrence determines the magnitude of risk.

10.6   Conclusion

Various success factors have been proposed for DES projects. This chapter has gone some way towards enhancing our understanding of the success of a DES project from the perspective of project executives. This study has highlighted the value of risk analysis for the successfulness of DES projects. The present study was undertaken to take into account the risks to support DES project in healthcare. The results of this research support the idea that the application of project management techniques can reduce the risk of DES project failure in healthcare. Furthermore, this research revealed that the use of risk management approach can handle the uncertainties related to the dynamic aspect of healthcare system and thus choose the best and most reliable scenario as one of the main outputs of DES project.

Even though the project team becomes sure that a model of simulation is valid, but this is no guarantee that stakeholders use it without being convinced of its benefits and its limitations [6, 15]. The proposed method in this study enables simulation experts to take into account the probable conditions affecting modeled system and convince stakeholders and decision-makers to use DES results by engaging them in sensitivity analysis. The application of risk analysis techniques in healthcare simulation projects is still in its infancy and there is a long way to go. It is hoped that the findings of this research will suggest an adequate level of interest for more applications, more tools, and additional studies to promote usefulness of risk analysis techniques in healthcare simulation projects.

A greater focus on managerial techniques in healthcare simulation projects can produce interesting findings for simulation software companies and management consulting groups in healthcare. Another possible area of future research in the successfulness of DES project would be to investigate in organizing workflow, planning, and the feasibility study of healthcare simulation.

References

1. Hamrock, E., Paige, K., Parks, J., Scheulen, J., & Levin, S. (2013). Discrete event simulation for healthcare organizations: a tool for decision making. Journal of Healthcare Management, 58(2), 110–124.
2. Roberts, S. D. (2011, December). Tutorial on the simulation of healthcare systems. In Proceedings of the winter simulation conference (pp. 1408–1419). Winter Simulation Conference.
3. Gul, M., & Guneri, A. F. (2015). Simulation modelling of a patient surge in an emergency department under disaster conditions. Croatian Operational Research Review, 6(2), 429–443.
4. Kolker, A. (2011). Healthcare Management Engineering: What Does this Fancy Term Really Mean?: the Use of Operations Management Methodology for Quantitative Decision-making in Healthcare Settings. Springer Science & Business Media.
5. Wilson JCT (1981). Implementation of computer simulation projects in the health care. J Opl Res Soc. Oxford: Sep 1981. 32(9): 825.
6. Gunal, M. M., & Pidd, M. (2010). Discrete event simulation for performance modelling in the health care: a review of the literature. Journal of Simulation, 4(1), 42–51.
7. Tako, A. A., & Robinson, S. (2015). Is simulation in health different?. Journal of the Operational Research Society, 66(4), 602–614.
8. Steins, K. (2017). Towards increased use of discrete-event simulation for hospital resource planning (Doctoral dissertation, Linköping University Electronic Press).
9. Musselman, K. J. (1994, December). Guidelines for simulation project success. In Proceedings of the 26th conference on Winter simulation (pp. 88–95). Society for Computer Simulation International.
10. Law, A. M., & McComas, M. G. (1991). Secrets of successful simulation studies. In Proceedings of the 23rd conference on Winter simulation (pp. 21–27). IEEE Computer Society.
11. Ghanes, K., Jouini, O., Jemai, Z., Wargon, M., Hellmann, R., Thomas, V., & Koole, G. (2014, December). A comprehensive simulation modeling of an emergency department: A case study for simulation optimization of staffing levels. In Simulation Conference (WSC), 2014 Winter (pp. 1421–1432). IEEE.
12. Gosavi, A., Cudney, E. A., Murray, S. L., & Masek, C. M. (2016). Analysis of clinic layouts and patient-centered procedural innovations using discrete-event simulation. Engineering Management Journal, 28(3), 134–144.

13. Swisher, J. R., Jacobson, S. H., Jun, J. B., & Balci, O. (2001). Modeling and analyzing a physician clinic environment using discrete-event (visual) simulation. Computers & operations research, 28(2), 105–125.
14. Zeigler, B.P., Traoré M.K., Zacharewicz G., & Duboz R. (2018). Value-based Learning Healthcare Systems: Integrative Modeling and Simulation Architecture. London: The Institution of Engineering and Technology.
15. Barnes, C. D., Quiason, J. L., Benson, C., & McGuiness, D. (1997, December). Success stories in simulation in health care. In Winter Simulation Conference (pp. 1280–1285)
16. Baldwin, L. P., Eldabi, T., & Paul, R. J. (2004). Simulation in healthcare management: a soft approach (MAPIU). Simulation Modelling Practice and Theory, 12(7-8), 541–557.
17. Jahangirian, M., Taylor, S. J., Young, T., & Robinson, S. (2017). Key performance indicators for successful simulation projects. Journal of the Operational Research Society, 68(7), 747–765.
18. McHaney, R., White, D., & Heilman, G. E. (2002). Simulation project success and failure: Survey findings. Simulation & Gaming, 33(1), 49–66.
19. Robinson, S., & Pidd, M. (1998). Provider and customer expectations of successful simulation projects. Journal of the Operational Research Society, 49(3), 200–209.
20. Harper, P. R., & Pitt, M. A. (2004). On the challenges of healthcare modelling and a proposed project life cycle for successful implementation. Journal of the Operational Research Society, 657–661.
21. McHaney, R., & Cronan, T. P. (2000). Toward an empirical understanding of computer simulation implementation success. Information & management, 37(3), 135–151.
22. McHaney, R. (2009). Understanding computer simulation. Bookboon. http://bookboon.com/en/understanding-ebook
23. Jahangirian, M., Taylor, S. J., Eatock, J., Stergioulas, L. K., & Taylor, P. M. (2015). Causal study of low stakeholder engagement in healthcare simulation projects. Journal of the Operational Research Society, 66(3), 369–379.
24. Brailsford, S. C., Bolt, T., Connell, C., Klein, J. H., & Patel, B. (2009, December). Stakeholder engagement in health care simulation. In Proceedings of the 2009 Winter Simulation Conference (WSC) (pp. 1840–1849). IEEE.
25. Jahangirian, M., Naseer, A., Stergioulas, L., Young, T., Eldabi, T., Brailsford, S., ... & Harper, P. (2012). Simulation in health-care: lessons from other sectors. Operational Research, 12(1), 45–55.
26. Lehaney, B., Clarke, S., & Kogetsidis, H. (1998). Simulating hospital patient flows. The journal of the Royal Society for the Promotion of Health, 118(4), 213–216.
27. Lane, D. C., Monefeldt, C., & Husemann, E. (2003). Client involvement in simulation model building: hints and insights from a case study in a London hospital. Health care management science, 6(2), 105–116.
28. Schmeiser, B. W. (2001). Some myths and common errors in simulation experiments. In Simulation Conference, 2001. Proceedings of the Winter (Vol. 1, pp. 39–46). IEEE.
29. Robinson, S. (1997, December). Simulation model verification and validation: increasing the users' confidence. In Proceedings of the 29th conference on Winter simulation (pp. 53–59). IEEE Computer Society.
30. Belson, D. (2013). Managing a Patient Flow Improvement Project. In Patient Flow (pp. 507–527). Springer, Boston, MA.
31. Lucas HC, Ginzberg MJ, Schulz RL (1990). Information Systems Engineering managers Implementation: Testing a Structural Model. Ablex Publishing Corp: Norwood, NJ.
32. Jones, C. (1994) Assessment and Control of Software Risks, Yourdon: Englewood Cliffs, NJ.
33. Ropponen, J., & Lyytinen, K. (1997). Can software risk management improve system development: an exploratory study. European Journal of Information Systems Engineering managers, 6(1), 41–50.
34. Wallace, L., Keil, M., & Rai, A. (2004). Understanding software project risk: a cluster analysis, Information & Management, 42(1), 115–125.

35. Dey, P. K., Kinch, J., & Ogunlana, S. O. (2007). Managing risk in software development projects: a case study. Industrial Management & Data Systems Engineering managers, 107(2), 284–303.
36. Bannerman, P. L. (2008). Risk and risk management in software projects: A reassessment. Journal of Systems Engineering managers and Software, 81(12), 2118–2133.
37. Sharma, A., Sengupta S., and Gupta A. (2011) Exploring Risk Dimensions in Indian Software Industry, Project Management Journal, 42(5), 78–91.
38. Vrhovec, S. L., Hovelja, T., Vavpotič, D., & Krisper, M. (2015). Diagnosing organizational risks in software projects: Stakeholder resistance. International journal of project management, 33(6), 1262–1273.
39. ISO Guide 73:2009 - Risk management – Vocabulary, International Organization for Standardization.
40. Hillson, D (2009) Managing Risk in Projects Fundamentals of project management. UK. Gower Publishing.
41. Sargent, R. G. (2013). Verification and validation of simulation models. Journal of simulation, 7(1), 12–24.
42. Kreutzer, W. (1986). System Simulation – Programming Styles and Languages. New York: Addison Wesley.
43. Balci, O. and R. E. Nance. 1987. Simulation development environments: a research prototype. Journal of the Operational Research Society 38: 8, 753–763.
44. Banks, J. (Ed.). (1998). Handbook of simulation: principles, methodology, advances, applications, and practice. John Wiley & Sons.
45. Sharma, P. (2015). Discrete-event simulation. International journal of scientific & technology research, 4(4), 136–140.
46. Loper, M. L. (Ed.). (2015). Modeling and simulation in the systems Engineering managers engineering life cycle: core concepts and accompanying lectures. Springer.
47. Seila, A. F., & Brailsford, S. (2009). Opportunities and challenges in the health care simulation. In Advancing the Frontiers of Simulation (pp. 195–229). Springer US.
48. Caro, J. J, Getsios, D., & Möller, J. (2007). Regarding probabilistic analysis and computationally expensive models: necessary and required. Value in the health, 10(4), 317–318.
49. Maria, A. (1997). Introduction to modelling and simulation. In Proceedings of the 29th conference on Winter simulation (pp. 7–13). IEEE Computer Society.
50. PMI, A Guide to the Project Management Body of Knowledge (PMBoK® Guide), (5themergency department.), Newton Square, PA: Project Management Institute, 2013
51. Hollocks, B. W. (1995). The impact of simulation in manufacturing decision making. Control Engineering Practice, 3(1), 105–112.
52. Robinson, S. (2004). Simulation: the practice of model development and use. Chichester: Wiley.
53. Banks J and Gibson RR (1997). Don't simulate when: Ten rules for determining when simulation is not appropriate. IIE Solutions, September.
54. Lowery, J. C. (1998). Getting started in simulation in healthcare. In Simulation Conference Proceedings, 1998. Winter(Vol. 1, pp. 31–35). IEEE.
55. Gunal, M. M. (2012). A guide for building hospital simulation models. Health Systems Engineering managers, 1(1), 17–25.
56. Robinson, S. (1999, December). Three sources of simulation inaccuracy (and how to overcome them). In Proceedings of the 31st conference on Winter simulation: Simulation—a bridge to the future-Volume 2 (pp. 1701–1708). ACM.
57. Kelton, W. D. (1999). Designing simulation experiments. In Simulation Conference Proceedings, 1999 Winter (Vol. 1, pp. 33–38). IEEE.
58. Lyles, M. A., & Mitroff, I. I. (1980). Organizational problem formulation: An empirical study. Administrative Science Quarterly, 102–119.
59. Brailsford SC, Lattimer VA, Tarnaras P and Turnbull JC (2004). Emergency and on-demand health care: Model-ling a large complex system. J Opl Res Soc 55: 34–42.

60. Law, A. M. (2009). How to build valid and credible simulation models. In Simulation Conference (WSC), Proceedings of the 2009 Winter (pp. 24–33). IEEE.
61. Pace, D. K. (2004). Modeling and simulation verification and validation challenges. Johns Hopkins APL Technical Digest, 25(2), 163–172.
62. Duguay, C., & Chetouane, F. (2007). Modelling and improving emergency department systems Engineering managers using discrete event simulation. Simulation, 83(4), 311–320.
63. IEC/ISO 31010:2009, Risk management – Risk assessment techniques. International Organization for Standardization,
64. Koh, S. C. L., Saad, S. M., Ahmed, A., Kayis, B., & Amornsawadwatana, S. (2007). A review of techniques for risk management in projects. Benchmarking: An International Journal.
65. Ezell, B. C., Farr, J. V., & Wiese, I. (2000). Infrastructure risk analysis model. Journal of infrastructure systems, 6(3), 114–117.
66. Hong, E. S., Lee, I. M., Shin, H. S., Nam, S. W., & Kong, J. S. (2009). Quantitative risk evaluation based on event tree analysis technique: application to the design of shield TBM. Tunnelling and Underground Space Technology, 24(3), 269–277. HIH
67. Kopach-Konrad, R., Lawley, M., Criswell, M., Hasan, I., Chakraborty, S., Pekny, J., & Doebbeling, B. N. (2007). Applying systems engineering principles in improving health care delivery. Journal of general internal medicine, 22(3), 431–437.
68. Tong, L. I., Chen, C. C., & Wang, C. H. (2007). Optimization of multi-response processes using the VIKOR method. The International Journal of Advanced Manufacturing Technology, 31(11), 1049–1057.
69. Gharahighehi, A., Kheirkhah, A. S., Bagheri, A., & Rashidi, E. (2016). Improving performances of the emergency department using discrete event simulation, DEA and the MADM methods. Digital Health, 2, 2055207616664619.

Printed in the United States
by Baker & Taylor Publisher Services